This work is copyrighted in its entirety. Any use or distribution is not permitted without the written consent of the author and publisher. This applies in particular to duplications, translations, microfilming and storage and processing in electronic systems. All rights reserved.

The prayers in this book do not replace the care of a physician, medical practitioner or psychiatrist, if there is suspicion of serious health problems.

The information in this book is presented to the best of the author's knowledge and belief. The author and publisher assume no liability for damages of any kind which may arise directly or indirectly from the application of the book or out of the prayers.

Copyright © 2016 by Anton Styger, Styger-Verlag

Internet: www.antonstyger.ch   www.styger-verlag.ch
E-mail: anton@styger-verlag.ch
English first edition 2013

Translation: Steven Lindgren

Graphic design: Mimmo Dutli
www.holisticharmony.ch

Cover photo: Pixelio

Cover design: Sonia Anderle

Printing: www.lulu.com

ISBN: 9781326498566   and   **978-1-326-49856-6**

ANTON STYGER

PRAYERS for the SOUL

Invocations
Prayers
Ceremonies for Redemption
and Liberation

Styger-Verlag

Contents

Preface 8

1. Understanding God 10
1.1 Morning Prayer 21
1.2 Daily Prayer 23
1.3 Prayer for the Night 24

2. Love and Joy Prayers 25
Love, Eternal Being, Jesus, Radiance of Love ... 26-28
Prayer Before Meals, Water Blessing 28-30

3. Releasing Earth-Bound Souls 31

4. Clearance of Astral Beings 35
4.1 Prayer for the Clearance of Astral Beings 37
4.2 Purification of the Aura 39
Short Clearance Prayer 40

5. Astral Garbage 40
5.1 Releasing Astral Energy
(Attached to Humans or Animals) 43
5.2 Clearance Prayer for Astral Energies
in Your Own Body, Outdoors 45
5.3 Clearance of Astral Energies from Buildings,
Homes or Objects 46

6. Forgiveness Ritual for Former Offenders 47

7. Resolution of Karma and Elementals 49
7.1 Release from Vows, Oaths, Promises
and Agreements 53
7.2 Release of Envy, Hatred, Anger,
Jealousy and Feelings of Pain 54

7.3 Clearance of Black Magic Thoughts,
Activities, Spells and Curses 54

8. Clearance of Individual Elemental Energies 55
8.1 Prayer for Healing and Clearance 60
8.2 Release from Poverty, Hunger,
Scarcity and Problems 61

**9. Clearance of Patterns of Suffering and
Elementals that Work Against Your Own Body** ... 64
9.1 Heartache 64
9.2 Clearance and Recovery 69
9.3 Prayer for Resolution of Fears 70

10. Body Meditation 71
10.1 Greeting Your Body Every Morning 71
Morning Ritual 73
10.2 Body Meditation to Strengthen and
Cleanse the Body's Lower Consciousness 74

11. Indigo Children and Their Parents 75
11.1 Crystalline Chakra Light Meditation
of Archangel Michael 78
11.2 Prayer for Indigo Children 83
11.3 Prayer for Parents 83

12. Words to Mother Earth 84
12.1 Prayer to Mother Earth 85
12.2 Prayer for Nature in the Spring 88
12.3 Prayer with and for the Sun 89
Prayer to the Sunrise 92
12.4 Prayer for All Creatures 92

**13. Harmonisation of the Nature Spirits
(Dwarfs, Gnomes, Elves, Fairies)** 94
Prayer for Troubled Nature Spirits 96

14. Animal Souls 98
14.1 Prayer of Release for Animal Souls. 99
14.2 Prayer to the Animal Souls. 100
Prayer to the Animals 101

**15. Purification from
the Lowest Earth Energies** 102

16. Wealth and Poverty 104
16.1 Suffering Due To Obesity 105
Ritual of Liberation from Obesity. 106
16.2 The New Era 108
Prayer for Abundance. 109

17. Wishes and Requests. 109
17.1 Wishes and Requests. 110
Example 1: A New Job 110
Example 2: Desired Partner 113
17.2 The Joy 116
For a Man – Prayer for His Beloved 116
For a Woman - Prayer for Her Beloved. 117

18. Suffering and Pain of the Body 118
18.1 Deprogramming Body Suffering,
Sickness and Pain. 121
18.2 Relief for Allergies from Pollen, Trees,
Shrubs, Flowers and Grasses 122

19. Casualties of War and Suffering Victims 125
Prayer for the Release of Wartime Experiences 126

20. Birth and Death 127
20.1 Baptism Ritual 127-129
20.2 Dying Process, Funeral Prayer
and Burial Rite 130

Death-Release Prayer 131
Prayer for Funeral Rites 134
Prayer for Ash Scattering 136

21. Liberation and Clearance of Demonic Burdens and Energies 136
Liberation and Clearance 139

22. Extraterrestrial Energies and Attachments .. 140
Clearance Prayer for Manipulation
by Extraterrestrials 141

23. Implants From Previous Lives 142
Prayer for Liberation from Old Diseases 143

Epilogue 145

Ritual from Archangel Gabriel 146

Preface

Dear Readers,

I am very pleased that you have chosen this book. By popular request, I have summarised all the instructions and prayers from the previously published series **'Experiences with the Other Dimensions'** (Band 1-3) and condensed them into a book small enough for a bedside table or even a handbag.

The huge number of responses from readers of the first two volumes have touched me deeply. It makes me especially happy to hear, that in these times my prayers have such a great appeal. Countless readers have already confirmed that these works have freed them to have a completely new understanding of God, and even helped them back to experience the joy of life.

In today's fast paced world many people feel empty inside, disoriented, and overextended. Many hear a call from within their hearts for a new direction in life. This call also strains apparently good relationships and creates big challenges. It is not the daily job or family alone that causes us to feel squeezed. No, it is the call of the Higher Self for liberation from the old baggage of past lives.

Many of us do not feel like we are at the top of the pyramid of our own positive experiences. Much more it seems like there is an inverted pyramid with its tip directly pushing down on people. And it is a pyramid filled, unbeknownst to us, with all the old, unkind impressions, experiences and actions from our past lives.

For many, this burden causes depression and illness. But most people never dream that their burden, inner turmoil, illness or misfortune has something to do with their past lives. Conventional thinking holds to the view that there is only one journey in this world from birth to death. But for those who question, it is interesting to observe how their body or their subconscious mind reacts to traumatic experiences that they themselves have not had in this life so far. Examples are inexplicable fears of enclosed spaces, heights, fear of fire or water.

Everyone is free to find his or her own meaning, and that is how it will remain. I would like to recommend you begin your own search too. This however means that we must face our dark side, and accept the immature actions of our previous lives. Only then can we forgive ourselves and clear the burdens we carry. The search for the origin of our problems outside of us is usually futile. The cause is almost always within us and needs to be discovered and cured. The root cause of most evil often lays incredibly many lives back.

Accepting who you are, allowing yourself to be forgiven, and loving yourself are the most important things for the future for us all.

Equally important is reverence for nature and the diversity of creation within it. The Earth would be fine without us, but we cannot be without it. Just as we see all that is great and divine in her, let us also celebrate the divine magnificence that is within us.

Anton Styger, springtime 2016

1. Understanding God, Prayers and Invocations

In various religions, prayer is an absolute must. For many people prayers are something that they have learned, but never really understood. They say them, only because they are familiar. They have simply not yet understood that praying for something can be wonderful and beautiful, if it is coming from the heart.

Some religions force their followers into a faith in which only the one God may be worshiped, otherwise he will be angry at the same believers who are in need, and seeking dialogue with him. There must be method to this madness, for it has succeeded to subdue the masses in various churches over the millennia (!) and beyond. Who does not know the fear of the wrath of God when the orders of the priests and the like are not obeyed? In this way, we immediately fall from God's grace and can never be happy in this life again. These are the facts: This is how people are manipulated and made compliant. The only way to avoid this fall from grace, according to the priests (who are representatives for institutionalised belief systems) is to pray at least so many Lord's Prayers, or make offerings and sacrifices. Offerings in the form of fruits or animals have been and continue to be a tradition in many cultures, even in Europe. But today's churches and institutions prefer money as an offering. That is the basis of their system. They cash-in several times for the same thing. It still works like this today, because many believers feel more comfortable and somehow relieved after making an offering. It feels like an obligation has been fulfilled, and a service done for God. It represents an easing of their conscience. In so doing, the individual 'believer' (ambiguous morality or hypocrisy) need not concern himself. As to where the money goes, obviously there is something good in it, because it pleases the treasurer of the institutions and fills up their coffers. Indeed this is even more beneficial, when the coffers are not constantly full.

But seriously, do you really believe that God needs, wants or demands sacrifice in whatever form? Everything already belongs to him. He's the creator of all that! Don't you think that he already knows when someone reluctantly gives something just to show other people that they are willing to donate? Interestingly, some people complain loudly about the wealth of politicians, managers, bankers or super-rich in their area. I often hear in conversations about God, "You know, I don't believe in a God. Just look at how so and so enriches himself by taking advantage of others. But he is still lucky; he amasses wealth without effort, is never sick and probably will get very old. Where is the justice? My wife and I work six or seven days a week, and yet earn only enough for small pleasures. Hell is here, and the devil is in heaven. No, I no longer believe in a God."

To this I like to reply: "Yes, this planet will be a hell if many people think about it the way you do." The more you feel resentment, envy or hatred towards the rich, the more you pull yourself to the opposite energy. And at the same time, you create Elemental entities that bind themselves to you. You are condemning wealth. You will become ever more poor and embittered, if you do not allow yourself and others be rich. Better to enjoy the success and wealth of others. Also, the rich need your prayers more than your curses. Look at what Jesus had said about this: **"the rich people who have no heart for the poor and suffering are the pigs of the earth. They will never sit at the table of the Father. Unless they show remorse and give everything they have taken back to the needy. Otherwise they will return to this earth as animals, so they can learn what helplessness and injustice are like."** But Jesus also said, and this is completely misunderstood by many: **"Those who have much will be given even more. Those who have little will have everything taken away."** (The Great Gospel of John, Jakob Lorber)

This has a purely spiritual meaning. Those who have come to terms with their soul's issues and with their own spirituality will be given increasingly higher levels of spiritual awareness. Those who are spiritually poor, and limit themselves to the material, will have everything taken away on the spiritual level. Everything, including the belief in a real-life god will be taken away, because they have not used the many possibilities that were available to them. The world is full of people who, after a natural disaster, famine or war, lose their faith in God. Few care to understand, that these events are based on the laws God gave to the material world, including the ones on our planet. People themselves put in motion the energies of the powerful spirit guide of the primordial elements. And these events, natural disasters etc., are consistently manifested according to the law of cause and effect within the material world.

What is lacking is an understanding of God and his work in all that is. Even more fundamentally, the understanding of what God is. He / She / It is only love. God is an immense universal love, without any expectations. In simple terms, it can be compared to the sun, which illuminates everything equally. It shines on the so-called good and the so-called evil. The sun itself does not make a distinction, even if we do not give thanks. It is our own reflection that makes the distinction. It seems the sun shines more often on those who love it and call to it. Only by radiating divine love on all that he has created, is everything real for us. God keeps all things alive as long as he wants to. He holds together the universal particles, out of which all matter is created. Without the will of the Creator, our planet would dissolve within seconds, and not even a particle of dust would be left. Only we humans are convinced that matter, such as stone, is without question stable, and will always remain so. But that is a mistake.

This corresponds to the illusion in which we humans find ourselves. All matter is, in truth, composed of vibrations, which are alive and emanate from continuously vibrating points of light. So there is not a bit of matter that is firm or completely dead. In many of the experiences that I describe, the downfall of others is often only the illusion of matter. Why? Because they cannot give up the material world, or even their own thought constructs. They cannot let go. Fear of the perceived void comes from ignorance. Most people are not aware that they do not need the very qualities, which make them unhappy, such as jealousy, pride, fear and anger. They do not know that they could change the cause of these, if they would just let go of their thought constructs. If someone says he has trouble with the statement: "Love God above everything", he has probably already asked several questions previously: "What is God? Does he expect something from me? Should I go to church every week? How can I serve him, and will he help me when I'm in trouble?"

These are all good questions, but let's look at them from the practical side. Imagine you are a parent. How do you handle your children in the following examples: You are in the kitchen. In the cupboard is a box of sweets that your children would very much like to have. The older child now comes to you, stands in front of you, takes a piece of paper from his pocket and reads you his request. Without emotion, he requests to have your chocolate. In the end, the child says three times: "I am a sinner and not worthy to get this chocolate."

What goes on inside you as you hear this? Do you not think, "Now she is completely crazy?" Or are you annoyed that this request was empty and just read off thoughtlessly? Would you feel like shaking her awake? The second child comes into the kitchen. She is cute and knows exactly how to deal with you in order to get what she wants, and even gives you joy while she does it. She comes to you, gives you a big hug and says, "Mommy, I love you so much, I would like to have the chocolate and really appreciate you giving it to me."

How does it feel now? Which child gives you more joy? Of course it's the one that loves you. Perhaps you feel anger towards the first child, or maybe you find it insulting, or even a nuisance. It is just an example to ask you, how God our father and creator would react to most of our requests and standard prayers. Do you think he would be the least bit happy? Not at all, and neither would all the other higher beings. They find these repugnant. Prefabricated prayers reeled off, pure lip service without thought and feeling, these must be as bad to God as fast food. But he does listen to every word that we send out from a pure heart with our true intention. God is not unattainable or far away, no, he is in us. And he knows every wish and desire that we have. He is everything that 'I am', all the particles and each molecule, 'I am' completely made out of him. No one is above me, except God, but equally no one is beneath me. Each of us comes from him; we all have the same origin. Thus, we are all of equal value to him. Indeed we are unique, free individuals, but we are all brothers and sisters that come from God. And we are all created from one and the same substance. There are people who say to me, "God still did not help me, even after I prayed for his help. That is why I no longer believe he exists."

To this I like to answer with a parable, as Jesus did: Suppose one of your children behaved unkindly to you for years, and abused you constantly as his parents. Once he is old enough, he sets off on his own. He breaks off contact with you and by so doing gives you even more pain. Your other children are always with you, you give them your affection, and you enjoy them in your life. Time goes by quickly, and perhaps in ten or twenty years, after you have come to terms with the situation, your prodigal son calls you and says, "Hello Father, I live in the U.S. and urgently need $30,000.

Please send it to me immediately, or they will put me in prison." What is going on inside your head? You have some savings and have planned to give it to the other two children so they can buy a flat. Would you be ready to help this son? Do you not think that you would probably give the other two a bad feeling, as they would then receive nothing? I can imagine that with a heavy heart, you would reject your son's plea, because you feel he is exploiting you. And until this time you could never make it right with him. So you say, it is his own fault for getting into this situation, and you tell him "No".

Thus is approximately the same relation that God and his host of angels have to us. Everything is reciprocal, and no one has to do anything unless they want to. We humans have received our free will from God. But if I have a love or friendship, and do not show care and appreciation, that love or friendship will become cooler and fade away. A friend, whom I had really forgotten about, would not be inclined to lend me money, if I asked him twenty years after we last met. A true friend, with whom I have maintained a friendship over the years, would come without hesitation to help me, because he knows that I love him.

Many mothers ask me: "How should I pray with my children. Should I pray to the angels or to God?" That in my opinion is just like thanking the housekeeper or cleaning lady for everything that you have received in your life from your parents. The fact is that every angel should be regarded as a servant of God. Our dear angels are the dearest and most faithful servants of our Lord (Father-Mother-God). They are the purest divine forms of love and wisdom. I usually see two angels close to a person standing at his side. God assigns them to each person, to give us individual assistance. But depending on the degree of friendship and love of an individual towards them, some angels appear bigger and more brightly coloured than others. Some are nowhere to be found, as they have separated themselves from their assigned person. They keep a distance when forced by a person's will.

And yet, when we pray, we should communicate directly to the Supreme Creator in us and to Jesus Christ, the Son of God, who stands ever so close. Nevertheless, many have easier access to their angels, due to lack of understanding of God. We must, of course never forget that our angels are our most loyal friends. They love us overall, as we people love our children: unconditionally. If we suffer, they also suffer. However, they usually cannot intervene, unfortunately. Why? Because we so often just want it or have 'ordered' it. But they also function as a filter and cause requests that we have, or have not just sent to the universe, be cancelled. Why? Because individual orders are frequently detrimental to our own life plan, or could even destroy us. We humans, unfortunately, do not know who we are as the soul that lies in our body knows.

"What does God expect of me?" That question is posed to me often, and certainly most of us have considered this key question. Well, I would almost say: nothing. Why? Think again from the human perspective. What do you expect from your children or your parents from you? Presumably you do not want to destroy the lives of your children with your unhealthy ego thinking.

You really only want your children to have happiness and health. You want the best for them. It is our greatest joy to see how our young ones become big, strong, independent and joyful. Then originality grows as they create their own image of life that must not match ours. If it makes our children happy, then we are satisfied. The same wishful thinking relates to the partner of your child. Maybe not your type, but the main thing is that he/she is happy.

You see, the less you inject your own interests, judgments, or even wishful thinking into your child's life, the more you become light and liberated. You are at the point of living with unconditional love and celebrating it. So this is how it actually should be: the children should be liberated. They should be released from the ideas that we have learned or experienced. These only have meaning for us, if at all, but not for our grown-up children. They are independent adults–souls with their own plan of life, and full of energy to find themselves anew. I believe, that this is surely the way God 'thinks' about all his children: real and absolutely not abstract, but complete love. But I assume that he cannot be happy about everyone, because he must be oppressed by the many that are blind and suffering. This is just my private opinion, and I may be wrong. Be brave, and find your own answer to this question. In this kind of conversation about God, I am then often asked: "Do you believe in reincarnation or do we live only once?" Jesus said: "People are made in the image of God." We have a spiritual heritage via our soul that is immortal, indestructible and eternal. God can and wants to learn through us. He delights in our creativity and development (if indeed there is some). Only arrested development leads to our spiritual death. We can develop ourselves in any direction, because we are completely free and unrestricted in the experience of this life and ourselves. God has given us the true fool's freedom. Of course, we can and must bear the consequences; otherwise our freedom would have no meaning. If we are to savour the impact of our thoughts and actions completely, we cannot assume that we are here only once on this earth. There is no cleaning lady to take away the rubbish of accumulated Karma and Elemental energies from our past lives.

If we only lived once, what kind of dirt would we take with us into the beyond, which would contaminate our soul's clothing for eternity? It is true that our Father forgives us for everything, so logically he could by his grace be the janitor. But what would we learn from that? How could we then develop, if our Father would constantly bear the consequences, due to his unimaginable, unconditional, complete love? Also, what would happen to all the innumerable bound souls and Astral beings, if we had only one life? They are quite real and they would remain in the Astral plane, or even lower, until they have looked at the effect of the many failings of their past lives. With clear foresight, God has designed everything for balance, experience and development, and has planned wisely. In the Swiss-German speaking regions we know the expression: "Wenn du das nicht verstehen willst, dann kommst du schon wieder auf die Welt." which is to say, "If you do not want to understand it, then you'll come back to this world." I find it interesting, and it makes me glad that we have this saying, although in our country, reincarnation is not represented in the religion.

During many lectures, I am asked: "How can I pray properly so that it actually works?" This depends on the closeness to God that you bring into your prayers, your honesty and the quality of your heart. The reading off of any given text, or pleading and begging will not work. Why not? Via his son, God has informed us many times, that we are not beggars, but rather his children. If someone begs to him, he will appear to be deaf. In a prayer, you give thanks or you make a wish for someone or for yourself. If what you wish for is not for the benefit of all concerned, then you'll be lucky if you were not heard. When I pray, I first feel God's compassion, so that I can address him with all my love. It feels just like when my own father held me and comforted me when I was a child. Then I thank him for allowing me all that I already have. And I thank him for his love, while sending my love back. In this way, even many of the greatest wishes have come true. Of course, they only came true if they themselves believed in it. Jesus said, **"Your faith has helped you"** and also: **"Become like little children, and everything will be possible for you. In this way, you can do more than I have ever done."** In the latter he referred to the time when we have obtained full Christ-consciousness, and have completed all of our self-imposed tasks and have gotten rid of all our bad habits. However I look into the world and its inhabitants, I can only think that this will take a long time. But God has time. He has created it for us, because what we call time does not exist in the hereafter.

The following prayers will help you to manifest new and powerful ideas. They are not begging prayers; on the contrary they affirm your divine self. They do not work merely by reading. No, you should be present with all of your thoughts and senses. Be inspired by the prayers, and create for yourself appreciations and invocations, released by your feeling of love. So you can awaken the divine within yourself.

1.1 Morning Prayer

Father-Mother-God, beloved source of all that is, I call you. I call you, my Saviour Jesus Christ, beloved, the greatest teacher and healer, God's son from the purest source of light. Thank you for knowing me and loving me. When I think of you, my heart becomes warm and I can feel your love and connectedness. I call you all, you beloved angels of light, joy and divine wisdom. Thank you for your loving embraces and support. God, I thank you for this beautiful light and this joyful day. Today everything is going very smoothly, because I know that you are always with me and I can feel your loving energy. Also, because I'm sure you are all within me and all around me, and I therefore am a part of you, inseparable since the beginning.

Beloved divine spirit, guide and protect me from mistakes and disappointments during the day and at night. My Creator let my inner energy centres and my aura shine like the pure divine light of dawn. Light in all, assist all people to open their hearts to your love. Thank you that because of your love in me, I can recognise and clear away all my old habits, judgments and suffering that my fellow man shows me as in a mirror.

Thank you that I am now able to resolve, in this life, all the old burdens of my unconscious past and be free for all time. Beloved Divine Energy, I am ever so grateful that I am permitted to live as a sovereign being and creature on this paradise earth. I am the 'I Am' in the here and now. I love you all. Amen

The following prayer is especially powerful, because it expresses the desire to give and the ability to give. Our idea of creation, combined with our feeling of love, creates glorious, vibrant energy and becomes a huge gift to our fellow human beings. In addition, we discover that our own great strength lies in our giving. We can give, if we want to. Those who can give will also receive, because the energy of wealth flows in both directions. Practice this visualisation. Create a picture in your imagination of the person to whom you want to send love. Imagine them surrounded in wonderful, colourful love light that is simply intense. It works because in this moment you are making use of the channel of divine consciousness. As a matter of principle, avoid words such as "I beg you, Father, that my child is not sick", or "I beg that I will be healthy." Rather, affirm in the positive, using the present tense that your child is perfectly healthy and intact. Give thanks also for yourself, for your health, and you say, "I'm healthy!"

1.2 Daily Prayer

Beloved Father-Mother-God, the source of all that is, I call you. I call you, Jesus Christ, beloved and greatest teacher and healer of humanity, Gods son from the purest light. I am grateful that you know me and love me. When I think about you, my heart chakra gets warm, and I feel your love and connection. I call you all, you beloved angels of light, joy and divine wisdom. Thank you for your loving embraces and support. I thank you, God, for this beautiful light, and for this joyful day. I'm grateful that everything is going well for me today. I am constantly aware of your presence with me, and I feel your loving energy.

I surround myself in your protective light that radiates in divine white, royal blue and divine gold colours (visualise). I thank you, Jesus, for the red-coloured light of love that shines from your heart directly into my heart. Your pink light of love envelops my aura and me. I would now like to send this pink aura and red heart light of love to others. (Visualise your heart radiating a pink flame of love that envelops your loved ones until they are absolutely and completely illuminated in the light of love.)

I would like to send love energy to:
my children
my parents
my friends
my work colleagues
also to the dead
(Always call each by name)

I thank you God, and you beloved Jesus and all you beloved angels and beings of pure light. Illuminate me on my path today, so I can walk with joy and light. So be it. I am the 'I Am' in the here and now. I thank you for my abundance, wealth and joy. Amen

1.3 Prayer for the Night

This prayer is about purification. We free ourselves every evening from the Elementals we created, unnoticed, during the day. The covering of violet coloured protective light, is most beneficial when you visualise it. Of course you use the support of your angels with this until you have it mastered. You are able to create protection and love around you and thereby sleep peacefully.

Night prayer

I call you Creator-Father Mother-God, source of love, with your human-god, son Jesus. I want to thank you for all that I was allowed to experience today, and for all I was allowed to perceive. I want to thank you, that I am perfectly healthy and intact, and my body is the expression of your divine love. I call you, dear angels of God's pure love, truth and wisdom. I thank you for the support and unlimited love, which flows out of you always.
I pray to you, great angel of the violet rays of light, envelop me with your powerful violet light. Your healing light wipes out my negative energy, thoughts and actions

from today, which I regret. I dismiss them into the light, releasing them from my field. Please spread your divine violet light over the whole room, throughout the house, over the surrounding land and across the whole neighbourhood. Your divine light protects me and my loved ones during the night from energies that would hurt us. I thank you, dear Angel of Light. Bring me in the level of your light, where my soul can recover and re-energise. I thank and I love you, beloved divine beings. I sleep calmly and deeply, until the morning, thank you. Amen

2. Love and Joy Prayers

Now I would like to offer you a few short prayers, and encourage you to let your own prayers flow from your heart. When I am 'overcome', completely spontaneously, then I know God's love is inspiring me. Unfortunately, many people dare not speak directly with the divine in themselves. But that is just what I want to encourage you to do, and indeed I would like to ask you to do so. Our Creator God does not want to be prayed to in the human form of Jesus Christ. Even his own mother Mary said that she does not appreciate to be honoured in this way. She explains that by being honoured in the human form a cult emerges, and people forget that she is only the mother of a God-man. To be sure, she has a pure spirit, but Jesus, her son, was the highest in the human form. The origin of Jesus as Christ is the spirit of God in Jesus.

True prayer is not worship, but a conscious relationship to God. Only then can I truly recognise myself. A quotation from the Gospel of John as transmitted to Jakob Lorber in the 19th century (Bd.9, p.51) should make this clear. Jesus told his followers in Jericho:

"Is then the pure will of God in man somewhat less than a divine will in God himself? And is man also somewhat less independently powerful, because God's will is at work everywhere, including in humans? Therefore shall a true human also come to be as perfect as is the Father in heaven? Is man not that, is he not also full of wisdom, power and strength?"

Here, dear reader, we see again the proof that we are much more than we anticipated. From this positive feeling of self-worth, we can give back our gratitude and love energy, and let these flow out. For if you let love flow out of you, you receive multiples of it back in return.

Every prayer is a dialogue with the highest source of love. At times it is easier, at other times harder to say a prayer, depending on how we feel. Therefore, each one of these four prayers may be chosen as an 'entry code' to our heart.

Love

Dear God of light and love in me and in all that is, I call you and thank you that I'm here on this earth at this time. Thank you that I am healthy in body, mind and soul. How wonderful it is to feel your pure love in my heart. How nice it is that I am sensitive enough, that I can and will share your love. What would I be without feelings of love, to myself, to the people around me, to the animals and to the plants? Merely an empty shell. Thank you for sparing me from this pain. Thank you that I'm not just a part of the whole, but a conscious part of you and your immense universal love. I am in love, and in you. Amen

Eternal Being

Dear Creator-Father Mother-God in me. I call to you. Thank you for this enlightening and wonderful day. I can see, hear and feel, how your influence is in me and in everything. I enjoy the beauty and diversity of plants and animals. Thank you for this life in this body, but still much more for the eternal life of my soul with you in the bliss of your abundant sky. I love you. Amen

Jesus

Dear Saviour Jesus Christ, I call you. Thanks for the great love that you are in my heart. When I think of you or call you, my heart warms. Then I know that you are always with me. For this I thank you, and ask you for your guidance. Help me not to deceive myself and allow myself to be seduced. Give me the clarity and courage to take the best path towards the goal of eternal inner peace and divine truth. Amen

Radiance of Love

Dear Father-Mother-God in me, I call you with all your angels of light and love. I know that your radiance of love is indescribably great and strong. Your light beings are the expression of the proof of your streams of life. I feel your love energy in me and your angels around me. Thank you for this wonderful contact with the abundance of your love. Through this I feel I am continually connected with you, the ultimate source of being. Amen

Prayer Before Meals

Father-Mother-God, Creator of all that is, thank you for all that you give us. Thank you for the food from our dear Mother Earth, who nourishes us faithfully. Dear God, bless our food with your love light. As required: Thank you wonderful animal soul(s) that you have sacrificed yourself for us. Thank you that we always have enough food. This we also wish for all the other people on this earth. Amen

For Vegetarians

Father-Mother God, Creator of all that is, thank you for everything that you give us. Thanks for the luminous, colourful food from our dear Mother Earth, who nourishes us so faithfully. Dear God, bless our food with your love-light. We are grateful that we always have enough food. We wish this for all people on earth. Amen

Water Blessing

Our drinking water from the tap, unfortunately, loses the vibration energy it had at the wellspring by being stored and transported in tanks, reservoirs and pipelines. It loses, so to speak, its consciousness of life. But it is still much better than the pseudo-mineral water bought in bottles. The highly praised minerals are unfortunately left in the bottling factories. Or have you ever found a clearly visible calcium ring in your drinking glass? This cannot really exist, because no minerals are present. By means of your will and your own love, you can transform any water. The best of course is tap water, because it has not been 'murdered' as it were.

Instructions

When you fill a jar or bottle with water, speak loudly and clearly: ***"Dear water, I love you, I thank you, you are good for me."***

You will notice that bubbles form in the container after about half an hour. The smell of the water changes for the better. This short speech to the water has a visible and energetic effect, because your will and desire always overcome the material. They are coupled with the soul. There are, however, higher vibration levels than just your consciousness alone that change water and food. These are blessings with the assistance of the Divine Spirit. If you have already spoken to the water or food as before, then you can speak aloud the following:

Dear God and Creator of all, I ask for your increased light and your blessing. Change this water according to how you have envisaged it, so that it is full of life and the best and keeps people healthy and even cures them. Thank you for your love for me. Amen

3. Releasing Earth-Bound Souls from Houses, People and Animals

The most effective clearance is possible if we are familiar with the area where the trapped souls, troublemakers or other Elementals are found. These are usually areas, rooms etc. where peculiar smells, cold drafts, or feelings of uneasiness are noticed. They are also places in the house where you feel a cold chill run up your spine, or get goose bumps on your skin. You may notice these feelings whenever you go to these places, or only occasionally under certain circumstances, which you should take note of. It is only by means of your aura, or energy field, that you are able to perceive these sensations. Depending on the state of mind and emotional condition of the trapped soul, the feelings you have can range anywhere from love and comfort to severe anxiety. The vast majority of people have far too little sensitivity to be able to notice anything in these situations. It may sometimes be the case, that the whole room is filled with Astral beings, or that some kind of parasitical soul consciousness is in the aura or in the consciousness of an unsuspecting person.

Procedure 1

In the area of the disturbance, or with the affected person in the room, you must first make yourself absolutely sure that you can and will carry out the clearance. And in a loving and heartfelt way you must assure the fearful entities that your intention is to provide assistance, clarity and compassion. Although these entities often present themselves as tough and confident, they are usually unsure and full of anxiety. It should be self evident that we can only be successful towards those beings, which are invisible to us, with a strong mind and mental strength (force of will). A friendly, loving, emotional approach will fail in this case, as will any attempt to smoke them out! We need the support of the Divine, and the help of the Archangels and their light warriors, as well as that of our guardian angels and spirit guides. A time when it is no longer light outside is best. At the beginning you light one or more candles. Then position yourself with your back against a solid wall or sit on a firm, comfortable chair. Do not forget that each word that you speak must reflect your focused intention. Just reading the text is absolutely useless, as no one will take you seriously.

Releasing Earth-bound Souls

Beloved God who is in me and in all that is, the original source of love, light and joy, I call on you and on Jesus Christ with my soul, here and now for assistance.
I ask all the divine holy angels, archangels and spirit guides to come in, and above all you beloved Archangel Michael, guide of lost souls back to their home. Please fill this space (this house) with divine, sacred light in blue-white and gold colours, and with violet beams of protective light and with a violet shell of protection. Please fill this space (house) with brilliant, glittering

light from out of the ground and the surface of our Mother Earth. This protective light should fill everything, including us, so that there are no shadows or places where the dark may hide. We ask you, dear God, friend in us, let this light be always there to protect us and to help us. Dear Archangel Michael, take these lost souls in your hand now and lead them away from their miserable existence, and put an end to their painful ordeal. Help them to break away from their false ideas and habits. Liberate them from the fear of death and the unknown beyond.

Now I will speak to you, soul, or the many souls that are here without knowing what you do and what effect you have. Awake from your deep sleep. Today is the 3rd of July, 2011 (Give them the actual date).

You are dead people who remain on Earth, and exist in an in-between world without time, bound to this Earth or imprisoned here. You do not know that you have left us, and no longer live in the physical world. For still being here, you have no one else to thank but yourselves, and it is because of your conduct in the last life. Either you have not believed in an afterlife in glory, or you were told that it simply doesn't exist. Your own thought processes hold you firmly in this false reality, and prevent a further progression into the light and joy. You can go back to your soul at any time. You only have to let go of the thoughts that are preventing you. Dismiss all your feelings of guilt and fear. Forgive yourself for the first time in all your life! Forgive all those you have hated or mistreated. Ask for forgiveness from all those you have unjustly treated or tortured, no matter whether it was by thought, by words or by deeds. And not only in the past life, but do this for all previous lives. You are free. No one will ever punish you. Our loving God would never punish you. Feel your freedom. Go now with these bright and loving spirit guides. They will lead you away from your burdens, with which you are now still tormented, and due to which you torment us so. This is the eternal law of cause and effect. We forgive you. We liberate you. We pray for you and love you. Do not fear going into the light in the name of Jesus Christ, our greatest healer and brother. You are free. Feel how your heart grows light and overflows with warmth and joy.

In order to free yourself from your guilt, I still have an important task for you. Help other souls that are in trouble, with the same words and with pure love. Keep this feeling of happiness, and now I say good-bye to you. I thank you, divine angels of love and light energy. Above all, I thank you Universal Love, God, friend in us. Amen

Procedure 2

If it turns out that there is still someone in the house or on the person, the ritual should be repeated at an appropriate time when it is calm and quiet again. It can also happen that a released soul returns after a certain time, or had not actually gone. With this soul you should be a little more energetic and assertive. You can say, for example, when they are in the room or can be perceived: ***"Go now! Have you still not understood that you are dead? Just go now."***

This is more of a 'wearing down' practise to express that we simply do not want an uninvited guest to remain any longer. It could also well be that others have arrived, who have noticed that there are people living here with compassion and warmth that radiate love, and they would also like to be helped. Then accept this task with gratitude, and help them. However, give them your conditions by saying, for example: ***"Okay, I'll help, but today I have no time for it. You can come back on Saturday night. Keep yourselves calm and loving, and do not disturb my family and me."*** Remember that everything we give from the heart comes back to us even stronger.

4. Clearance of Astral Beings

Now in this practice, it often happens that after a clearance of earth-bound souls, a total relief and improvement comes in for the affected person. But after a certain period of time passes, the old misfortunes or circumstances can reappear. What has happened? Many times we do not know in what kind of situation the soul was in, or why it was Earth-bound. Unfortunately, it is usually the case that they were themselves victimised or possessed. When we now help these souls so that they can go, where do all the Astral beings go? Usually they stay in the area, having not understood what has happened. They are often only fragments of consciousness from the lower Astral plane that gather here in groups. One cannot always perceive them clearly, because their energy is not sufficient to form a fluid body. They usually take on the form or expression of grotesque faces and grotesque body parts. They are also often not fully awake and their state of consciousness might be that of a trance or sleepwalking. Depending on what sort of deep Astral plane they come from, they may react strongly to our thoughts and emotions. They feed on the possessed, so to speak, or are attracted to them. Also, our radiated Elemental energy unintentionally attracts them. Imagine yourself as a radio transmitter that constantly sends out your old patterns or misdeeds as behaviour Elementals into the ether. The only stupid thing is that you have no awareness of your transmission, which is running continuously. Yes, you do not even know who you were or what you did in your previous lives. Therefore it is absolutely necessary, and surely also enlightening, that you free yourself from these as quickly as possible. Therefore go and get texts on Karma and clearance of Elementals, and study them carefully, so that you can help yourself to feel free. Here we need more divine assistance, because alone we only have the will, and are, so to speak, just the trip organiser.

4.1 Prayer for the Clearance of Astral Beings

Beloved God who is in me and in all that is, the original source of love, light and joy, I call on you and on Jesus Christ with my soul, here and now for assistance. I ask all the divine holy angels, archangels and spirit guides to come in, and above all you beloved Archangel Michael, leader of lost souls helping them to their home. Please fill this space (this house) with divine, sacred light in blue-white and gold colours, and with violet beams of protective light and with a violet shell of protection. Please fill this space (house) with brilliant, glittering light from out of the ground and the surface of our Mother Earth. This protective light should fill everything, including us, so that there are no shadows or places where the dark may hide.

We ask you, dear God, friend in us, let this light be always there to protect us and to help us. Dear Archangel Michael, wake these soul fragments or Astral beings now, and translate for them what I have to say. Then take these lost beings in your hand and lead them away from their miserable existence, and painful suffering. Help them to break away from their misconceptions and rigid thought patterns. Free them from the fear of death and the unknown beyond.

Now, I speak to you, you many soul fragments of a past time that are here now: You are prisoners of your own thoughts and deeds, or of those who created you. You do not know what you are, what you do, or what effect you have. You are earthly negative energies of deceased persons, existing in an intermediate space without time, and living in the Astral world. You have probably not noticed that you're gone from us and you are no longer physically alive. You, who are here with other like-minded entities, going around on our planet, torturing many people: do you really know what you are doing? You plague and harass people, but all the harm that you do to others, just reinforces your own suffering and pain. Listen, immediately stop this and be released from your old thought patterns and habits. Know that in all the universes, the law of cause and effect can never be avoided. You have broken many laws and have not observed the divine principles. You have probably not loved yourselves, or have never even known love. Now, above all, whom should you love, or even understand? Forgive yourselves and all who you see as your adversaries or enemies. Learn to accept yourself and to love yourself. Go now with these dear light-beings, who guide you to a place that is better for you, and where you can rediscover yourself. There you will be informed and educated. Do not be afraid.

For still being here, you have no one else to thank but yourselves, and it is because of your conduct in the last life. Either you have not believed in an afterlife in glory, or you were told that it simply doesn't exist. Your own thought processes hold you firmly in this false reality, and prevent a further progression into the light and joy. You can go back to your soul at any time. You only have to let go of the thoughts that are preventing you. Dismiss all your feelings of guilt and fear. You are free. No one will ever punish you. Also our loving God would never punish you. Feel your freedom. Go now with these bright and loving spirit guides. They will lead you away from your burdens, with which you are now still tormented, and due to which you torment us so. This is the eternal law of cause and effect. We forgive you. We liberate you. We pray for you and love you. Do not fear going into the light in the name of Jesus Christ, our greatest healer and brother. You are free. Feel how your heart grows light and overflows with warmth and joy. Keep this feeling of happiness, and now I say good-bye to you.

I thank you, divine angels of love and light energy. Above all, I thank you Universal Love, God, friend in us. Amen

4.2 Purification of the Aura from Negative Energies

Actually, all of us feel it when something is around us that influences or changes us. In many cases a person is irritated, 'beside himself', with anger or frustration. But he is also often sad and depressed without even knowing exactly why. In all these cases, it is almost always an attack being carried out by negative astral energies. It can even manifest itself in the form of pain or a migraine.

Against this, I propose a brief clearance, so to speak, as a precautionary measure. It will not case harm in any case. **This is particularly effective when the attack was not longer than 24 hours ago. Please speak this out loud several times a day:**

Short Clearance Prayer for Astral Energies on One's Own Body

I call to you blessed Saviour Jesus Christ and you beloved Archangel Michael for help. Dear angel please come with your light sabre and cut all these negative energies away from me, which are such a burden for me. Please purify my auric fields as well as the solid cells of my body.

Thank you dear Saviour Jesus Christ, that I remain free, because you are the light in me. I love you. Thank you dear Archangel Michael, and thank you all my dear angels for your protection. Amen

5. Astral Garbage

These energies, which come from the Astral world, are the most common negative energies that are found attached to humans and animals. A non-clairvoyant, may well sense a heaviness about them, but they don't know what it is. I perceive them as dark clouds and even as grotesque faces and creatures in the aura of a person.

They arise from the low thoughts, and hence the negative aspects of living people and of course of the deceased. As such, it is unfortunately true, that these negative energies are always in abundance around us. As they develop from negative thoughts, words and deeds, they are, so to speak, 'subtle garbage'. Since it is energy of a negative kind, we must deal with it more severely than when getting rid of the souls of the dead. They attach themselves to us like sticky mud, and of course, they also do this to the dead (earth bound souls). Few people know, that the stronger and lower (more evil) they are, the more they are at the service of the great master of the dark. Astral beings are the hiding place for malice and deceit, and almost always harbingers and servants of the devil.

When dealing with such forms of energy, there is really just black or white, either or. That means I must always fight for the integrity and purity of my soul, and maintain my full conscious attention to all that I do. If I allow myself to spend a longer period of time with grey energies around me, I must not be surprised if I suddenly see only black from that day on. Actually, I should defend my soul, and thus my whole life all the time, like a housewife who would never, ever let a mangy, flea-ridden dog come into her house. Of course he will try again and again, but she would throw stones at or do whatever it takes. Most likely he would always come back into her vicinity, as soon as she turns her back. This parable is to make you understand how the Darkness works. The dark side knows our weaknesses better than we do. Those who give up just a little are at the mercy of the darkness, without having realised it. In this way, people are lulled to sleep, and easily separated from consciousness of their soul in this life. The rude awakening and with it the sadness of the separation, leads them to seek only after physical death.

It could happen to anyone of us. Also the media is used as a tool of darkness. Many people are seduced by the TV, lulled and separated from themselves. This is an open secret. The first contamination for young people happens by listening to the negative vibrations in certain music (I like to call it 'monster music'). It is a disgrace that concert organisers, such as the Eurovision contest, accept and promote such vampire garbage! But the 'responsible parties' are not even aware of their responsibilities. Perhaps they are already employees of the great, dark power? How many of the still developing unstable youth really have the power to defend themselves against the deliberately low vibrations that such music consumption exposes them to? They are especially vulnerable, because they observe that millions of others see it as a normal.

On this point we must be questioned as consumers, parents, but also politicians. Do we have the courage to liberate TV programs from violence and negative content? Our children will thank us.

When releasing people from this kind of negative attachment, we must be aware that they do not perceive us as having a bodily form, and therefore they cannot understand what we say. They are just clouds of negative energy, that are emitted telepathically. A telepathic thought, as we can call it, also does not know what or who it is, let alone where it comes from. But, it resonates with us, and that's how it influences us!

This means nothing other than, that we either carry this negative energy in us, or are already karmically familiar with it. This is the only way we can attract these Astral energies and be attacked by them.

Procedure

When I want to clear someone who has asked me for assistance, I stand close behind him and place my hands on his shoulders, or if he is lying down on his upper body. By so doing, the power and influence is enormous. This is because I am wrapping my pure intact aura over the other person's aura. The personal presence and proximity of the helper forces the lower energy to respond immediately. Then I speak the following prayer loudly and with conviction.

5.1 Releasing Astral Energy (Attached to Humans or Animals)

Beloved God, the Creator, and Jesus Christ in me and in all, I call you. And I call you; you angels of pure light and love. Make this room and the house pure and free. Illuminate all materials from the foundation up through the roof, so that nothing negative can nest or hide. I call you, dear Archangel Michael, please help us. Serve as a translator and lead all those negative energies away. I thank you for this.

(Speak from here with a really harsh voice like a command!)

Now I speak to you, you low frequency energy attached to this man. You are fragments, or even consciously programmed negative energies from people who have died or perhaps even from the living. You do not know who you are or where you come from. You originate from the dark side of people who have thought, spoken or behaved unkindly. That made you trash of the worst kind. What you are doing here is the opposite of love and is contrary to the divine law of the freedom of our souls. As an advocate, I speak out for the freedom of this human!

I will not have a discussion with you, because you come from the Darkness. Because I will not repay evil with evil, I urge you to leave, ... (name of the person being helped) *immediately! The Darkness, however, must not be allowed to become stronger as you return to your originators. Therefore I wrap you in intense love light that will make the Darkness weaker by your return. So I ask you dear Saviour Jesus for your divine light of love, so I can surround these dark clouds in it.*

Now visualise and feel how your warm, pink light of love comes out of your heart and flows completely over the person and their whole aura!

Then you end the clearance with the following words:

In the name of Jesus Christ I command you to disappear, and return to where you came from. You may never attach to anyone, neither person nor animal. Dear Creator-Father Mother-God, anchor and reinforce your divine light of love for (name of person), *to the extent that their entire body, all cells and all molecules, are cleansed of dirt and deposits of these negative energies. Make his aura strong and pure, so he will know if something negative approaches. Protect him so that he can achieve his life plan. Help ...* (name), *to carry out all duties and overcome all obstacles with grace and ease. For you alone, God, are the power, the light and love in us. I love you and thank you, Amen*

5.2 Clearance Prayer for Astral Energies in Your Own Body, Outdoors

Due to long periods of degradation and manipulation of their energy bodies by these negative energies, the will of many people is weakened or even broken. Often they are no longer in a position to read or speak a prayer. Then the only option is rebellion and shouting out in order to give oneself courage. These energies know us all too well, especially our weaknesses.

Outdoor Procedure
Go far enough from home, so no one can hear you. Call Jesus Christ for help and for his red light of love. Try to visualise that you are enveloped in the pulsating red light of Christ and that it embraces you fully.
Speak loudly:

Jesus Christ I ask you for help. Please envelope me with your vibrant love light and save me from these torturing energies. (You have to imagine these energies on your neck and back as a cloud.)

Now speak this very loudly and forcefully:
You are dirt energy, you have no business here, I don't want you! Disappear in the name of Jesus Christ! You disappear right now and I'm free! Thank you Saviour Jesus Christ, I am free. Amen

5.3 Clearance of Astral Energies from Buildings, Homes or Objects

Often whole parcels of land, the buildings situated thereon, or even an entire region is affected by astral energy (astral garbage). The energy of the lowest thoughts or energy from terrible suffering was simply left behind or 'built in'. It continues to have its effect forever, unless it is cleared. Without clearance and transformation, people who live in such polluted houses can never be happy.

To rid your house, property or region of such low level contamination, you need to visualise with the full power of your heart, that all the floors, walls, ceilings, furniture (everything that is matter) are wrapped and saturated in a beautiful light of Christ-love:

Beloved God, beloved Elders of Light, I need your help and the help of the beloved angels of light and pure love, to clear the dark burden from this material. Please wrap the land and the building from foundation up through to the roof in the light of love and wisdom, so that dark energy is no longer able to hide here. Help these forms of energy, return to their source. Make this house free from negative thoughts and feelings, and also from the money power and greed of the builders. Dear Saviour Jesus Christ, please fill all the material of this house with your pink-coloured light of love. (Visualise) *I set you free. You are free. So be it. Thank you beloved Creator and thank you, angels of light and joy. Amen*

6. Forgiveness Ritual for Former Offenders

When clearing earth-bound souls and Astral souls, it sometimes happens that they simply will not go away. I have also noticed this with souls that are not very strongly attached to their environment. I wondered why they would not go away, not even after several clearance prayers? Now I know why. They were former offenders who are eternally bound to their victims, as long as no one intervenes. For many people, I can see the former perpetrators as incarnated souls of animals. It sounds absurd, but this is a way of making an apology and atonement. Imagine that a woman's nasty neighbour suddenly becomes her lap dog, or the hard-hearted landlord becomes the mascot rabbit in their rabbit hutch, or the customers who never paid on time, now in the pigsty of a neighbouring village. I am serious all of these cases exist! Most of these suffering souls of offenders even come back to us as animals several times in succession.

But watch out. This is no reason for gloating or Schadenfreude. Perhaps we too have taken this difficult path at some point. It would be really sad if, by my own misconduct or by a gross misunderstanding, I would become the poodle of my wife's beauty salon. That's what I call 'Eigentor', or self-defeating folly.

I dare not think of the countless perpetrators of countless wars, concentration camps and all the bad things that have happened. Yet I see them again and again, hardly a stable or feed company, without souls of former offenders. It ties together the universal law of cause and effect, with the universal law of balance. The grace that we ourselves are capable of is therefore the letting go and forgiveness for these souls.

Prayer for Forgiveness and Pardon

Dear Creator-Father Mother-God, I call you for help. Dear Angel of the Purest Light, please support my efforts. I would like to help this soul that is bound to me, so it will be free.
Now, I speak to you, soul(s), which are around me. Listen carefully to what I have to tell you. You are bound to me, because in some way, in some previous life, you have done me wrong. I want to free you, because in this life it is easier to forgive something, of which I know nothing more. I would also not like to rehash the pain and torment that I have experienced. For me, you are gone forever. I forgive you, and let everything go. It is also stressful for me, when suffering beings know about me. Your sadness makes me unhappy.
Luminous angel, convey to those soul(s) my wish for their liberation from bondage to me. Beloved God, save this soul and also me from further mistakes and unnecessary suffering. Fill our hearts with your incomprehensible love and clarity. I thank you and your angels. Amen

7. Resolution of Karma and Elementals

The word karma comes from Sanskrit. In Buddhism, it is the form of re-incarnation a person receives in order to fulfil the destiny that was shaped by their previous actions and experiences. Karma means literally 'action', 'word' or 'deed'.

We people in the West don't know what to do with this term karma. Many even refuse to talk about it as a matter of principle. Surely because they believe it is associated with a strange religion, or otherwise need to explain it as something with mystical charm. I try to demystify this concept and describe it as: life experiences in past and present lives. Life experience is naturally made up of all experiences ever had, from the good to very bad and painful. These are therefore the basic patterns we carry within us.

Every thought every feeling and every action is registered. Imagine all the elements, not only for this life but from all lives lived through each incarnation are recorded exactly and stored on a huge hard disk or CD. God has this most gigantic of all the databases or, more precisely, he is it. Nothing is missing in it, not the tiniest bit. Even our very DNA structure stores the memory of all experiences.

The resulting feelings and impressions are what matter for us. If it is feelings of happiness and joy, then for sure they are not burdens. No, they would like to invite us to always continue to feel or experience them again and again. On the other hand, we have the vices of humanity. These are fundamentally not joyful. They plague and torment us. They point to an absence of love.

They consist of lovelessness, whether mere thoughts of harming another creature, or actually doing harm by word or deed. Just the idea of causing harm to someone can initiate an energy carrier a so called 'hollow globe' (The Gospel of John, Jakob Lorber), which accompanies us forever from that moment on. In each new life, in every new incarnation, it is available for us to make use of, thereby increasingly strengthening it. And all this happens without our being conscious of it. These hollow globes (initially only labelled carriers with no content) can be filled with more and more energy of a similar nature. As packets of energy that belong to us, they become our Elementals. Always located around us, they are filled with our feelings and fragments of consciousness. Actually, they are our tools for life, because they correspond to a huge library of past experience and knowledge.

Unfortunately, people tend to be creatures of habit. They follow familiar, well-trodden paths, as opposed to new and unknown ones, which pose the 'danger' of new knowledge and experience. That would be terrible! They could experience something new. Consequently, they always meet the same Elementals, and typically these correspond to their familiar vices and addictions. The burdens of ones life can be compared to a backpack filled with heavy stones that make the journey difficult. What does a clever person do in this circumstance? They try to make do with less and think, "What is really necessary for my journey, and what can I leave behind?"

People who have a good heart, and not just physically, take the time to listen to the quiet whisperings of their souls, and accept these as truth. For others, the connection between their heart and their conscience is buried. These are people who believe in the material, and feel separated from God and respond only to the material world around them. They have become deaf to the divine within that is themselves, in their hearts. Their resulting dissatisfaction or inner conflict is directed towards the outside. No wonder such people lose self-confidence, as more and more difficulties come into their lives. The constant idea of 'Not Enough' becomes a permanent state of 'Nothing Accomplished'. Soon these people begin their decline, and self-hatred grows. When self-love is not to be found, a downward spiral begins, which few can stop before their death. Then the next life begins already with a burden from the last. And so goes the cycle of re-incarnation and karma. For many people, there are no beautiful views in the period ahead.

But stop! Do not despair. Everyone can do something about it. Very much can be done. Disconnect and separate yourself from these patterns and energies, and take your head out of the noose that you have made for yourself. Whether you send out negative energy against yourself or another, both are equivalent 'offences'. When you condemn someone, you condemn his creator, therefore you condemn God. And it's no different for you. If you condemn yourself, you condemn your Creator. It's the same effect emanating from the same cause, the same basic evil.

When we realise this, we are perhaps aware of how important it is to love ourselves. We all have to learn to respect ourselves. Because when we respect ourselves, we automatically pay attention to and respect others. And that's really what we want from others: That they respect us as we are and let us be who we are. You are the expression of the divine, and have the perfect light in your heart and pure love. Believe in yourself, then you believe in God! Vices are permitted by God. They are components of the human path to knowledge, and they arise from our free will. They arise from our forgetfulness of who we are and what we are, and from our ignorance and unconsciousness of ourselves. Our vices cover over our luminous soul. And as our souls get encrusted by this filth, so are we prevented from experiencing our lives with lightness and joy. Since this is caused by our human desires, the soul gives up it's yearning to completely clear away the vices that we have chosen for ourselves. As these self-created Elementals are mostly made of old and strongly manifested energy, we can hardly release them with a simple prayer. It takes a little patience and perseverance (created over one or more lives), to dissipate the 'well-fed and cared for' Elementals that are our vices.

The best time to clear such an Elemental is just when a thought or a feeling arises, which questions whether the divine within you is being honoured. Stop and ask yourself the reason. Recognise that you have created a form of unkindness that you do not need anymore. Then wrap this, your own energetic entity, in the greatest possible sympathy, because this poor creature was indeed created by you. If you address your vices, and call them by name, you can even apologise to them. Give them your love, and ask Jesus Christ to heal this part of you.

If you recognise your own failings as a lack of love, then you're on the right track. If you can't name any of the concrete vices (those are directly addressed in the next chapter), which still give you suffering, you can work through the following list. It refers directly to these tough, old patterns from time immemorial. This can happen over a few intense weeks, as long as you still feel mentally unwell for indefinable reasons. Give each word weight and meaning, so they can work for you and liberate you sooner. Understand this as a conversation and giving of orders to both the sub and highest consciousnesses.

7.1 Release from Vows, Oaths, Promises and Agreements

I ask all angels and higher beings for assistance at this time. I hereby release myself from all vows, oaths, promises and agreements, which I have expressed in this life and in all previous incarnations. I ask all soul brothers and sisters, whom I have hurt or offended, for forgiveness and clemency. I forgive all who have bound and imprisoned me with promises, oaths, vows and agreements, now and at all times. I forgive myself for all the suffering that these commitments have caused for me and for others, now and in all other previous times. I'm free, now and in all following incarnations. I am the I am, in the here and now. I send all the resulting suffering now and forever into the divine light of the universe. I am free. I am free. I am free! Thank you, Divine Source! Amen

7.2 Release of Envy, Hatred, Anger, Jealousy and Feelings of Pain

I now release all feelings of envy, hate, anger, jealousy and pain that have been inflicted on me in this and in past lives. I forgive all of my tormentors forever. I ask all for forgiveness, and regret deeply that I have caused others to suffer from envy, hatred, anger, jealousy and pain, in this life and in all previous incarnations. I thank you, I love you, please let go of these negative feelings, now and forever I am free. I am free. I am free! Thank you, Divine Source! Amen

7.3 Clearance of Black Magic Thoughts, Activities, Spells and Curses

In the here and now, I forgive all for the black magic thoughts, activities, spells and curses they have been addressed to me, that have given me so much pain and suffering. This applies to all of this life, and to all prior time. I love you. I also ask all for forgiveness for the pain and suffering, that I have inflicted by means of black magic and malevolent thoughts. I love you. Please forgive me. I ask you, divine light workers and spirit guides to assist in releasing me and all involved from the things that have tormented us and prevented us from standing in the light of the universe. Now separate us from these negative Elementals that have arisen from experience, suffering, and recognition and are well disposed toward me. Please stand behind me and help me to push them away. But please leave the way open for new experiences in these new and fast-moving times. I love you. I thank you, my angel of light and you, God, original source of all love. I am free. I am free. I am free! Thank you divine source! Amen

8. Clearance of Individual Elemental Energies

A first way to release individual Elementals is by recognising and accepting their existence. They equate to our vices such as envy, resentment, greed, hatred, jealousy, revenge and lust for power. Unconsciously, we hold and maintain these energies in the lower vibrational frequencies over countless lives, and even cause them to grow by our thoughts, words and deeds. Even the seemingly weaker vices such as lying, malice, cynicism, Schadenfreude, arrogance, vanity, greed, etc. are accumulated in this way.

These are self-created, autonomous spirit beings, which exactly match our weaknesses and are created by us. They can grow to a considerable size and accompany us all the time. We are forced to deal with them. They exist because they have a special purpose. They help us to recognise that they have been created by ourselves, and that we should get rid of those patterns on our own. Sometimes they plague us so severely that we develop a deep desire to be free of them. Then we have already made a big step in our personal process of maturation.

If we do not notice however that they torment us, and that we should change our own thought patterns (they do not yet torment us enough…) then they keep coming back to us. They then work like a radio station with only one song, always playing the same piece of music, which becomes all too familiar. Due to this effect, it happens that our own Elemental energy attracts and brings in more of the same kind of energy from the Astral world. This leads to a permanent increase!

The repertoire of our vices is immense. Since we do not even remember exactly what we have created for ourselves by our thought patterns as children and young people, we can only guess what bad habits we have developed in previous lives. But rest assured, they are all still there. Of course, not all thought patterns are equally powerful. But a huge number of them carry destructive potential. Due to the Law of Polarity, each idea creates a dual that is a counterpart. This is our compensation resulting from the absolute balance that must exist in the universe. So we are theoretically always free to decide, which part of the energy we want to activate.

The best example is love. From the absence of love it is possible to stimulate the opposite of it, hate, the most destructive emotional force. It is possible that people who don't feel loved or do not love others can develop hatred against themselves or against former loved ones. Let's look at exactly what happens: Say a person says "I hate you", this act creates energy with the low frequency vibration of hatred that is addressed to someone. This energy, which is newly called into being, will remain active for all time. And, unfortunately, is often called in anew, that is used, unconsciously by 'normal' human thoughts and actions. In this way, the energy becomes increasingly more powerful and conscious, that is, it becomes an Elemental being.

Our physical death does not erase this energy. It exists until we consciously rid ourselves of it. It is always around us, even when we are dead. With some people, the Elemental beings keep tapping them on the shoulder again and again, saying: "Hello, here I am", according to the motto "I'm still here". This only happens to those who are developing in a different direction than their Elemental, which was created by them at an earlier stage. So if they now suddenly go lovingly and deliberately through their life, they will be surprised from time to time by unexpected feelings of envy, anger, hatred or jealousy. It can be safely assume that these are their previously created Elemental energies.

Phobias are also in this category. Have you ever wondered why a young person might have some fears that in your opinion have no reason or explanation? For example, they may have fear of water, fire, spiders, confined spaces or high altitudes.

Some people have even created their Elementals with such great power and autonomy that they believe there are strange powers or actual demons around them. These boundless energies with partial awareness then act like a horde of wild devils that torment their creators.

They cannot be cleared so easily, because they are made out of malice, deceit, seduction and deception. These people have created a masterpiece of the dark side for themselves that is full of treachery.

Of course, these Elemental beings want to continue their existence for all time. With their partial awareness, they feel very important. Ultimately, their roots go back to the dawn of mankind. They are enormously important for our spiritual development, as we can eventually recognise that we no longer need this energy now and for our future.

All addictions are consequences of our vices. Our soul is the 'seeker'. It is looking for truth and liberation. But when our limited human consciousness does not want to seek, it finds ways and means to be blind and deaf. In the aura of all addicts, I clearly see the presence of Elemental beings, which torment those affected very much. So I can well understand that addicts try to befuddle themselves, so as not to take a close look at precisely what is needed.

Due to the worldwide spread and increased proliferation of addictions, you can well imagine that these people easily attract to themselves Astral beings with exactly the same addictions as their own. These in turn absolutely do not want 'their addict', that is to say their victim, to stop listening to their own addiction Elementals. By satisfying the addiction of the affected victim, the Astral beings are further satisfied. The reason it is so immensely difficult to break free from an addiction is that the particular influences of all foreign energies must be taken into account.

The vehement Search of the Soul shows up, however, in completely different addictive behaviours, than our well-known drinking, smoking, drugs and gambling. These are not always recognised as such in our society. For example, with motion-seeking addicted people, the problem is in running away from responsibility to their soul. Most people think that lots of jogging or bicycling is good for their health. I personally feel that any sport that is associated with obligations on the body is more destructive than constructive.

My clairvoyance allows me to see Elemental energy, and in the aura of almost all athletes, I see negative Elemental energies. Maybe you are surprised? I see some athletes who would rather be destroyed by constant strain, than pay attention to anything else. Subconsciously, they seem to know that there is much within that is dark and unpleasant. But just because it is ignored, does not mean it goes away. Those who do not face their problems, help them to get bigger. For all who do not vibrate freely, who are not completely sound and full of love, strive for understanding, rehabilitation and redemption.

If you have decided to resolve a particular vice or addiction, then please treat it as a being that you yourself have created. Then wait for a time when the addiction is clearly evident in you, because then this Elemental being is in your immediate vicinity and directly addressable. Try to heal the Elemental you created, in the midst of your acute stress and the energy it creates, with the following prayer.

8.1 Prayer for Healing and Clearance of Individual (Self-Created) Elemental Entities

I call you, my Creator-Father Mother-God, of infinite wisdom and light, who is in everything that exists. And I call you, our dear brother Jesus Christ, with all your angels. I ask for help and healing.

Through lack of understanding and ignorance I have created a thought pattern that has become increasingly stronger and bigger. It has been hurting me a long time, perhaps several lives. Now I am ready to resolve it and to be free. Beloved divine power; please help me here at this time. Now I call you, my fear (or addiction, anger, pain, saying the exact name). *Come to me please, because you are a part of me that I now recognise. I am very sorry that I have created a being with such a low frequency of vibration. As the child of God that I am, I have acted in ignorance. Please forgive me.*

Now sit down very close to me on my lap, come near to my heart, like a child of mine. You know, I have created you, certainly a long time ago, and I no longer know why. I always notice your presence, but I never knew what it meant or what you wanted to tell me. Here and now, I take responsibility for you and set you free. I will ask for your healing, so you can walk in love and light. I release you from your task and thank you.

Dear Saviour, pray for us in the here and now and surround us in the light of your love, so we both are cured as one unit. Please dissolve the energy created by me in your light and in your love. (Let a few minutes pass so that your love can do its work.) *Thank you God Almighty! Thank you, Divine Source! Amen*

8.2 Release from Poverty, Hunger, Experiences of Scarcity

Many people, even in our northern latitudes, were starved in one or more earlier lives. From the knowledge or rationality perspective, this is obscure to them, but is expressed through their body. They carry these dramatic details in their cellular memory. For their entire lives, these people have problems with their weight and their appearance. They also have constant fear that there is too little food, or that tomorrow there will be nothing else to buy. They often keep the refrigerator hopelessly overloaded, and must regularly throw away most of it. They suffer because of their excess weight and appearance, and hate what is reflected in the mirror. Most believe that it is simply their hormones or glands that are not functioning properly. In order to dampen their frustration, they reward themselves with constant sweets and other goodies. This obviously leads to too many calories and produces the current problem.

Quite a few people that appear thin have the same problem, because they want to fill out, but cannot do it. They may even be admired by many, but they would like to change themselves or otherwise be fixed. They have kept themselves resistant to the starvation they experienced. The steadfast people are burdened by their will to survive their former starvation, which is quite normal.

What does a starving person think about? About food, full of bacon and beans and a table with sweets. Who would be thinking about clearing negative thought patterns of scarcity well before death, especially if you have gone a month without having something to eat? So the gluttony pattern is preserved and relived in every new life. When this drama plays out together with feelings and emotions, it seems even more intense. During those times not only individuals suffered from hunger, but entire families, the village, a region or the whole country. How that must have been like, to have dug a grave for someone in your family, perhaps for your own child, knowing that in a few days others would be digging one for you. Something like that must be terrible, and it happened to many, and over a few years gap again to countless others.

Clearance of this Disease Pattern

Actually, these Elemental energies cannot be cleared or dissolved in the same way, since the emotional body is additionally affected by stored experiences and suffering, and the physical body has stored commands as well. Clearance is not easy when done alone, because our mind bucks and blocks. It has problems with this, and will not or cannot understand. It was also not present in that body. Done with a couple (your husband/wife/girlfriend) it's easier because you can repeat and not be held back by your ego. This would also work with a group clearance.

I call you, you angels of love and truth, please help me with the clearance of my old patterns of suffering. Now I call the sufferings that were created in the **super- and**

***subconscious levels of my mind** when I died of hunger in past lives. I created you, but I do not want or need you anymore. In this life there is no famine and no shortage of food. You patterns of suffering and dying from hunger and want, I liberate you, and let you go. At the same time, I clear myself from you, too, you patterns of fear of abundance and excess. You are all free and I am free. Beloved Angels, please help me with this.*

*Now I call the pain and suffering, the sadness and the fears of loss in my **feelings and emotional bodies**. You originated a long time ago from the suffering of famine, and I also do not want you anymore. I dismiss you. You are free, as I also will be free. Beloved Angels, please help me with this.*

*Now I call my **body**, and all of you, the countless many spirits of my body. But above all the **senior guiding spirits** of my organs. You have suffered with me during the time when starvation and death were very common. I ask you to release all these energies and memories, but also those related to excess and abundance. I currently suffer from obesity, as do you. It is in the past. I want you to remember the source of your being, of perfect health and healing. Release everything now that goes against this. Clear away all that I have commanded to you out of ignorance in this life and in the distant past. Be free! We will remain healthy together until the last hour. I love you because you serve me with all your love, and not only in this life.*

I thank my God and Creator for my body. It is wonderful and perfect, and the expression of the divine. Please help me to fix these problems. My body and I are now as you gave to me. Amen

9. Clearance of Patterns of Suffering, and Elementals that Work Against Your Own Body

9.1 Heartache

Have you ever noticed, that most people blame others for their own failures and misfortunes? This way of thinking is most clearly evident when someone is sick. Often the reason is attributed to their hard work, or to their colleagues or superiors. Who is already aware, that they have developed this way through patterns of thought that hamper them in their progress, self-confidence or the joy of life? Such reductive thinking patterns are not necessarily created in this life. More often they are brought along and possibly strengthened. How easy it is to create a vicious circle without knowing it: when all is not going well for people, they tell themselves they don't feel quite right, and they let their condition decline further. Over time they don't acknowledge that anything is going well. Eventually it feels right to be inferior. Soon such people actually become incapable of almost anything. That's interesting: By their own destructive evaluation system used to criticise others, sooner or later they begin to criticise themselves. From excessive self-criticism they can begin insidiously, to find themselves repulsive and later even to hate themselves. Should such a person feel better for some reason, they will of course forget the feelings of hate that they had against themselves. However, these are all periods of time saved and anchored in the lower consciousness forever. How many people have in their vocabulary words such as "I will/can no longer see, smell, hear or feel that!" Countless times a day this is spoken or thought. No one realises that these are direct orders to our bodies. But it is so: the 'chiefs' in our bodies are lively spirits of nature, who respond to the commands of human consciousness. Fortunately for us, they do not usually shut down an organ or bodily function promptly. The shutting down is usually preceded by a prolonged period of degradation of the corresponding organ.

Hence we do have enough time to work on counteracting these developments and on revising our negative thought patterns. But unconscious humanity does not know and do this. If the body would respond more quickly or immediately, these commands would be devastating, and perhaps more recognised as such. The fact is that these commands are confirmed for a long time again and again, even over several lives before they are finally fully executed.

Not infrequently, doctors contribute to derailing people's health unnecessarily in this way, but more often it is the patient himself, in sheer desperation. In this way it is possible for health issues or pain appear, for no clear reason.

Let me give four examples of health problems that were created, where none were supposed to be:

• A teacher lost his eyesight within two months. One of his favourite expressions was: "I can not see that anymore." Our body is composed of innumerable, directed natural elements which we can call nature spirits. That is to say, each cell has its completely specified task, which it was created to carry out. Under normal conditions, our body works perfectly, but beyond this it is unable to do anything else. Under no circumstances can our body understand our more or less complicated intellectual achievements. No, in this case its simple interpretation was: "I will go blind." With a change in his verbal expressions and especially in his thought patterns, after a short time the teacher's vision returned.

• A mother, who had big trouble with their teenage daughter, suffered suddenly from a painful breast infection. She had five big lumps. The gynaecologist advised for immediate amputation, as he diagnosed cancer. The woman had continuously said: "I failed as a mother." After she internalized the text below, the inflammation disappeared within a week.

• A woman who had been paralysed in two prior lives hated and despised her life in each. In this life she came into the world healthy, but often had the feeling that from the waist down something was wrong. She often felt a dull pain and numbness there. No doctor, no expert could ever find anything concrete. In time she began to condemn and hate herself. This created an increasingly clear disability. Soon she no longer believed that she would ever recover. Yet, after clearing these thought patterns with prayer and meditation, her condition became stable. After a few weeks, her feeling that she would be paralysed completely disappeared.

• A man suffered for years with psoriasis all over his body. It got worse, and when I met him, even the head and face were heavily infected. He looked really awful, and he suffered so like a wounded animal. He had started to despise himself, and would not go out in public. To me he commented that he was a victim of magic or voodoo curses. I had to enlighten him that he had initially received his psoriasis only to offset a situation from a previous life. But then he had compounded his suffering with more negative thoughts directed against him. He did not believe me. Shortly afterwards a medium told him that in a previous life he had mocked and scorned others with this disease. For this reason he had to endure it himself as compensation. Now he was ready to work with my kind of prayers. He did this so intensely that he freed himself of his psoriasis completely after two months. You could not see anything on him. It was as if he had never had the disease.

In addition, I notice that many women, who are suffering, are acting out old behaviour patterns that are directed against their own bodies. They carry these in themselves over many lifetimes. In a spiritual sense these patterns look like this: that they are not capable, not fully satisfied, do not have their own opinion, or they have been created just to serve. It's no wonder. The deeply rooted thought patterns in the female sex result from what was done to them for centuries. In many countries, the oppression of women still continues openly. Also, the dictatorship of the churches and religions was and continues to be with us. For centuries, countless women have been emotionally crippled in this way and today, many lives later, still suffer from this. Millions of wise women, expert in the healing arts, were hung on the accusation of witchcraft or heresy. 'Man' had the right to use cruelty to get rid of them, and even did this in the name of God. That such evil pigs, as Jesus called these tormentors, would become poor animals, I can well understand. Unfortunately, it is still true today, that the Catholic Church for example, considers childbearing to be the only role for women. Female leadership in the religious community is not open to debate for them.

It is also becoming apparent that many women today have fears about **becoming pregnant**. Not a few women experience powerlessness and even fear for their lives, every time they allow a man to come in. These fears took root in times long before the clarity and openness that we have today on sexual themes and contraception. Such deeply seated fears have led to an ever-stronger rejection of the feminine in the body, and it must be clear to everyone, that self-destructive tendencies are the consequence. These fears and destructive thought patterns (elementals), never disappear by themselves, especially when the women that create them don't want to know about them. I personally declare that this explains the obvious extreme high rate of diseases of the female sex organs. Countless women, for whom I was able to make these connections, have cleared themselves of the old entanglements with the help of prayers. Believe me, almost all confirmed to me that the great suffering they carried for many years, was blown away in a short time.

After this short list of real life experiences, you can now hopefully imagine, that a soul is often suffering and constrained, without any obvious cause. For this I have developed the following prayer, which will help to heal bodily suffering.

Please do this type of prayer with intensity and perseverance, because most of the negative thoughts directed against us have been repeated countless times at all times. Many human souls have regularly cursed and damned themselves in so doing. This prayer will undo all that. You yourself have the power to do it! You have the power to do this now! Everyone has the right to life and healing.

9.2 Clearance and Recovery from Patterns of Suffering in Your Own Body

Mistress divine angel of healing and of pure love, I call to you for help. I would now like to clear away all the thoughts and behaviour patterns that I direct towards my body, which cause me unnecessary pain and suffering. Thoughts that I am not healthy, or will have the same illness that my parents have, or other such thoughts and worries. I am now resolved to clear these patterns of imperfection and oppression against you, my beloved body from this and all previous lives.

I call you, Spirit of my body, and I apologise to you. I have unconsciously thought and done many harmful things, and you have always obeyed my orders often to my sorrow. Now I know that this suffering is unnecessary, and I beg you to let go of all that means suffering to me. You are the expression of the divine and of perfection. Your wholeness is sacred. Wholeness is healing through love. Without you as a healthy vessel, I, Soul, cannot fully experience the world. I pray to you, spirit of my body, to remain neutral, so the divine Universal Love can work its healing and transformation.

Now I ask you, Source of All That Is, for your divine golden light to dissolve and transform the toxins and deposits as well as all the rays of antennas and phones, into the most marvellous light. So you, my dear body, are healthy and strong again. I love you, and therefore also you Creator, as the grandest and most glorious love energy. Beloved Angel and Friends of Light help me do this. So be it. Thank you!

9.3 Prayer for Resolution of Fears from the Subconscious of the Body

Such fears may be, for example the following:
- Fear of one's own body
- Afraid of one's own feelings
- Fear of misunderstandings in relationships
- Fear of being unsuccessful in a relationship
- Afraid of not being loved

Beloved Creator, Source of all that is, please help me to relax and to heal my body. Beloved spirit of my body, I embrace you and your feelings. I love you, for you are my expression of the divine and of perfection.

Forgive me if I have suppressed you with my feelings and fears. You are free, dear body! We will start over fresh, one more time, in the light. Forgive me for the misunderstandings that I experienced or attracted as a child or in relationships. I am now clearing these thought patterns and emotions, and releasing myself from them.

I had great fear that I could not do enough for others, and as such could also not do enough for myself. For this arose my fear, that I could not make and keep good relationships with others. Forgive me and disentangle this knot, to clear all these fears!
I am counting on you!
I love you! You are whole and complete. You have a healing reaction to loving feelings.
You are fully capable of love, and thereby make my partner and me happy. You are attractive and have a wonderful aura. I thank you that you're so perfect. I love you.
I love you, and that's the most important thing. We are one and always happy, because we are as a unit independent and always free. Let everything go, what prevents you from being loved or from giving freely. Forgive me when you get the feeling that you are not adequate. Clear away all of these thought patterns now! Thank you!

10. Body Meditation

10.1 Greeting Your Body Every Morning

The vast majority of people do not realise that they make negative commands against themselves and constantly weaken themselves. Even little children see their parents or uncles and aunts suffering when they are sick. This is especially so for the older generation, grandparents and their siblings. But also we see other old people sick and suffering. Naturally we think this must be the way it is when you are old. The logical conclusion: I will also have to be like that when I'm older.

This information is already stored within small children. Actually, it doesn't much matter how old one is when we start holding such restrictive thought patterns, because it is always the same soul. Who remembers once they are twenty years old, what information they saved at the age of 5? But they are always within us, if we do not undo these '**commands**'. Thus for many, diseases and ailments appear out of nowhere that cannot be explained by doctors. These phantom phenomena or sufferings arise at precisely the time when we imagined we would be sick. On the contrary, we should erase all these 'orders'! It is worthwhile to do this every morning and simultaneously we strengthen ourselves mentally. Of course, I can say this as a human at the time of my death, because I certify to myself: I am healthy to the last hour.

The other is the fear of wasting away and having great suffering before death. So if I'm constantly afraid and do nothing to prevent it, I will also have to live through this. It is also not missing among the 'death scenarios' we observe in our relatives and acquaintances. Even if someone came back claiming it is not be so bad, we would not be ready to believe him, so stuck we are. But now be done with it.

We imagine ourselves in front of the mirror and see ourselves with gratitude, as the Creator created us. That doesn't bring satisfaction to countless people, even when they are beautifully built. Especially not those who are 20-50 pounds overweight. But it's so hard, because they are carrying the old beliefs: their body is sinful and bad. They despise themselves, you can hear the hatred in their words. There are experiences from many past lives, where they prepared their bodies for more suffering than joy. They were indescribably unhappy, perhaps forced to marry, and felt abused and used. This creates an aversion that goes as far as contempt. It's not only looks, even feelings and emotions are rejected, are frowned upon.

Watching in the mirror, you should touch and caress their body lovingly. Maybe it has never received this from you in a lifetime.

Morning Ritual

Dear body, I look at you, I love you. You are great, the expression of the divine. I am happy to be in you as a soul. I apologise for everything negative that I have spoken, thought and done to you in this life and in all previous lives. Forgive me please. Then you and I will be free. You are healthy until the last hour.
Today is a good and beautiful day, we are managing everything with ease and simplicity. Dear God, my Creator, I thank you for my beautiful and healthy body, I love you.

10.2 Body Meditation to Strengthen and Cleanse the Body's Lower Consciousness

If you do the following outdoors in the fresh air, this exercise will greatly strengthen and energise your body. It is important that we draw in light from our mother earth, who is called Gaia because she is a living being. Standing up straight, bring the light up from below, through your feet, up through the root chakra and perineum chakra, and into the body. At the same time we breathe fresh oxygen into our lungs. We bend the knees slightly, and bend the upper body slightly forward, to press the used air out from the lungs. While doing this you can declare aloud what you want to let go of. Then straighten up again, breathing in the fresh air audibly. You may allow your arms to open slightly.

Say this loudly or softly

Invocation of Mother Earth and Declaration of Intent

Beloved Mother Earth, Gaia, wonderful consciousness of our planet earth, I call you and ask for your wonderful light energy. Energise, cleanse and harmonise me with your beautiful vibrations.
I ask for your pure white light to rise up on the inside of my legs. In my pelvic area your light becomes yellow and is pink around the solar plexus. In the heart area, the light becomes red and then blue towards the shoulders. In the head area it becomes purple. Outside the body the light is green, as it slides back to Mother

Earth. With every breath I let my burdens go out of me. I say this out loud to my body.
Dear body, now release all forms of torment that come from the ideas which I myself or others have imposed on you. I send the glorious light to flow into you, and you will have purity and more freedom with every deep breath. I can feel the circulation of light and heat in and around my body.

In the end
Thank you Mother Earth for your energy and love. I love you and thank you for all that I am. For it is only through you, that I have a physical body and have food and drink every day. Hail to you.
Now, dear body, I thank you. I love you. You are my expression of the divine and of perfection. Without you I would be only a spirit and could not have this earth experience. So it is. Amen

11. Indigo Children and Their Parents

Dear parents, if you have noticed that your child told you things that you could not understand since the time they began speak, its possible you have an indigo child at home. They get this name, because the aura of such people as perceived by clairvoyants to be a gorgeous indigo blue. But there are also crystal and star children with different coloured auras. As you know, this kind of young person is a real challenge. Their statements and also their behaviour are different than other children. An Indigo child, at first sight, is not compatible with our existing intellect and family norms. They question. Indigos say most of what they feel, see and recognise in an open, simple and clear manner. Many adults struggle, because by themselves they had to take in all this knowledge. They feel the need to really learn, at home and, later more so in school. However, the learning content has always been presented to them this way, in order to reach their level if understanding. What's the big struggle? So here comes the knee-high boy and speaks of something that you yourself have never heard of. And even more, your child speaks constantly of what it sees or perceives somehow. It further talks and even plays with beings that in your eyes do not exist. You as parents often have the feeling that you cannot join in there, or label your child as one who has fantasies. It is possible that your child tells you stories of your childhood and adolescence, and you can no longer put them in the fantasy drawer. How is this possible? At this point, most parents are overwhelmed, and the worldview that they have built over many years begins to sway. Many parents of Indigo children note that they can learn from their three year old, I mean apart from what is shown us daily by our children. There are some stories about Indigo children in my other books **Experiences with the Other Dimensions**.

Unfortunately, many Indigo children are not at all happy in their situation, with their choice of life here on earth in a body or with their parents. They feel completely alien, and especially so in their cramped little bodies. Often Indigo souls are not anchored properly in their bodies (incarnated). They would prefer as soon as possible to be back home, back in the light. To understand your child better try to imagine that you are wedged in the body of a three year old as an adult. This will give an idea of the least some of the limitations the soul of your child has to go through. The other limitations are similar. In order to understand more precisely how it is for such a soul here on earth, you should imagine that a soul that is in the light enjoys unlimited freedom. They imagine something, and it's there. If they imagine themselves to be at a certain place, then at the same moment they are there. Your Indigo child still has the memory of this, at least when they are very young, and later perhaps only subconsciously. But now, here on earth, when these children would like to have something, they must wait, because on the material level things do not just get conjured up. What a change! Would you not be impatient if you knew that it really could go faster? The fact is that Indigo children do not come into this world with burdensome Elemental energies. They are free from karma. However, they are by no means immune from creating Elemental energies now in this life itself, nor from accepting them from others. I advise doing the **Crystalline Light Meditation** with Indigo children, to help them stabilise themselves within their physical bodies. Later, you can teach them a short form of it, which they can use by themselves at any time. The other prayers strengthen the understanding of Indigo children, and their parents, that being here on Earth is their self-chosen situation.

11.1 Crystalline Chakra Light Meditation of Archangel Michael

This special and powerful meditation brings the ability to cleanse the chakras, strengthen them and connect them with each other. Many people are not even aware that they have chakras and how they work. The name 'chakra' comes from Asia, and roughly means light cup, light lotus or light centre. The chakras connect the external light energy of the body's Aura with the physical body. As spirit energy is present in the aura, as well as in the physical body, it is very important that all the connections are open to the light. If individual chakras are closed by feelings, emotions or pain, the body sometimes gets out of balance. The aura is an energetic mold for the body. Without the aura there is no body. Unfortunately we see many people who have closed a number of chakras and suffer. In this meditation, we want to harmonise our chakras, as we imagine them individually and visualise the crystalline light. To make a simple understanding of the divine-crystalline light, let us imagine a Christmas miracle candle that turns in circles in our hand. A chakra looks like a funnel with the wide end pointing away from us, and the narrow end pointing inwards. For each chakra, we dwell on the light cone for 1-2 minutes that is we visualise it with our mind's eye. Only by our will and by our power of imagination, can we make it possible for the divine light to purify and strengthen us. We live on the planet of free will, and without our will there is nothing. For without our order or command, no angel-beings of light can intervene and help.

I prepare myself by making myself comfortable in a quiet room where I won't be disturbed, and begin to breath evenly and calmly:

Chakra - Light Meditation

Now I call you, Father-Mother-God, source of all that is, and all your holy light and archangels to me here in this room. I ask that this room and this house will be cleared of low thoughts and Elemental energies. Only pure light and emotions of joy and happiness have a place here. They strengthen me and give me joy. Now I call you, dear Archangel Michael, bearer of the pure crystalline divine light energy. I thank you for your presence and for your help. Please give me as much of this pure light, as my body can tolerate, and no more.

- We start with the **alpchakra** which is about 6 feet above your head and go through all the chakras individually, each working for at least 1-2 minutes.
  I ask you for the crystalline light.

- The next chakra is the **brow chakra**, also called the third eye. *I ask you for the crystalline light. Erase all holograms and false images.*

- The **palate** or **mouth chakra**, *I ask you for the crystalline light. Let out only words and phrases that come from the soul.*

- The **throat chakra**, *I ask you for the crystalline light. Erase all my suffering and memories from the past.*

- The next is the **violet divine centre** (This is in the upper chest under the collarbone). *I ask you for the crystalline light. Clear all negative elementals.*

- The following are the **nipple chakras**. *I ask you for the crystalline light. Clear all my old suffering, feelings of loss and the pain of motherhood.*

- The next is our **solar plexus chakra**, *I ask you for the crystalline light. Release all confining restrictions.*

- The **spleen chakra** (4 cm above the belly button) *I ask you for the crystalline light. Kindle the light of the spleen.*

- The **navel chakra**, *I ask you for the crystalline light. Erase all the commands from ancestors, as well as family karma and dramas from past lives.*

- The **hara chakra**, about 4 cm below the navel, which for women relates to the internal sex organs. *I ask you for the crystalline light. Clear all old suffering and memories from past lives.*

- The **sexual chakra**, *I ask you for the crystalline light. Erase all humiliations and sufferings from past lives, such as incest, abuse, rape and torture. Transform everything into gladness, joy and harmony.*

- The **dam** or **sacral chakra** (between the legs). *I ask you for the crystalline light. Make this Ring chakra full of light, so the channel of light can flow through.*

- The **knees**, *I ask you for the crystalline light.*

- The **ankle chakras** (tops of feet). *I ask you for the crystalline light. May the light glide through the feet.*

- The **soles of the feet chakras**, *I ask you for the crystalline light. At the same time I want to connect myself with the love energy of* **Mother Earth**, *for without her agreement, I would have no material substance. I greet you at this point, and thank you for everything and for every day that I may live on you and with you. I connect myself to you, to your holy light. Let me be in sync with your increasing frequencies, so that I can develop myself better and more easily. I thank you for your boundless love energy, and for the beauty that you give me. Dear Archangel Michael, open a light channel from my feet to the top, so that the crystalline light can go up, as in a tube, and connect at the top to the Alp-chakra.*

- The **back of the knee chakras**, *I ask you for the crystalline light.*

- The next one is my **root chakra** (tailbone, coccyx) *I ask you for the crystalline light. Erase all the injuries here from earlier times.*

- The **heart chakra**, *I ask you for the crystalline light that shines from behind through my heart to the front. It forms an illuminating tube, and arouses in me feelings of happiness and great love.*

- The **neck chakra**, and over both shoulders. *I ask you for the crystalline light. I want this light to tumble down my spine like glowing diamonds, until they reach my root chakra and then to flow up again* (I imagine this visually). *I wish that through this wonderful light, all my nerves, which emerge from my spinal cord, are fully illuminated and strengthened.*

- The **palm chakras**. *I ask you for the crystalline light. Open and fill my hand chakras, so I can heal myself or others.*

- The next is my **crown chakra**, *I ask you for the crystalline light.*
Dear Archangel Michael, please now connect all my front and back chakras with bands of this light. Fill my entire body and all my cells with divine light and warmth. I thank you dear Archangel Michael, for your love and work. I thank all the other angels of love for their closeness and undying support. I now take leave of you and my Father-Mother-God. Amen

When you work in a group, the sense of selective warming is stronger. When one leads and guides us during the meditation, our mind becomes peaceful, and we are able to enjoy our soul consciousness freely. After doing this prayer, you can do a daily prayer in this **short version**:

Dear Archangel Michael, I want to activate and fill all of my chakras with your crystalline light. I thank you for your help and love. Amen

11.2 Prayer for Indigo Children

My dear father, primal source of all, you gave me the opportunity to come to the planet Earth. I want to show people that the true being of the heart is pure love. I ask that I can have a warm feeling in my body, and use it as my vessel. I thank you Father, that you made my body perfectly safe and sound, so I can feel comfortable in it. I thank you that my parents liked me and accepted me with love and with great responsibility. I thank you, God, for the truth and that your love light can shine through me, and can illuminate the path also more clearly for my dear parents. Because of this, it is easier for them and for me to be in material form. I thank you for this life and the pleasure of it. So be it. Am I the I Am. Amen

11.3 Prayer for Parents

Beloved Father God, Lord of all, we thank you for giving us this beautiful gift of sunshine. We intend to exercise our responsibilities entirely, and offer this child a beautiful and secure place on this earth so we can share the joy of life.

Thank you, God, for the love that flows from this child. It reminds us of where we come from and of where we may go back to again. Help us to be led by the pure thoughts of our hearts, rather than by the distractions of the material-minded world. Thank you, God, that this pure essence reminds us of the lightness of our being, which is a great gift from you. Light and joy are in us and in everything. So be it. Amen

12. Words to Mother Earth

The vast majority of people living here on earth do not have any spiritual awareness. They are convinced that there is nothing spiritual. Others feel very religious and have a faith or belong to a church. They take what they are offered as real and true, but they are no longer open and searching. The idea has been instilled in them, that there is a God or, as in India, thousands of gods who are not here but far away. They have been told that the gods are not accessible unless an offering is made, or the priests are consulted. This is about the same as when a cow or a pig is made to believe, that without the master or the farmer, they would starve in the countryside. So I will explain how we limit ourselves, and how this has been done in so many lives before. If a person thinks in this way, it is about the same as if one assumes the world around us is all hard, stiff and dead, and then go for a walk, and shortly after a volcano erupts. This is fatalism. The earth moves and changes shape every time the moon raises or sets. Then these people would say again that this is just like a rubber ball, which is malleable. What such people cannot accept is the fact that we, as physical beings, are dependent on the earth, but she is not dependent on us. These people can only see with the greatest difficulty that we are actually spirit, but they can't see this for animals or plants, and certainly not our earth, the material. They may dare to think that an 'earth-worm' could be accompanied, or even controlled, by a spirit. However, the earth is such a large body to us that no one can really imagine it has a soul or is guided by its creator. And yet it is so and it even has a name. Her name is Gaia, and she is very feminine. Our mother earth shares her mission with the great 'spirits of matter'. We can talk with them and they react to us. If we suffer, so do they. They exist through us, and are part of us. If you are in open, untouched nature, try to call them, and you will perceive them. They are literally waiting for us, and all that is expressed with honesty and pure love is appreciated.

You can also call the spirit of the air, the clouds, the wind and the storms, and you will suddenly discover playful figures and shapes in the clouds, as you have never seen before. Or call the collective consciousness of the plants or animals, and you will be surprised by what you suddenly feel.

I would like to propose the following **prayer to Mother Earth**. It is an earth healing prayer or more like a conversation:

12.1 Prayer to Mother Earth

Dear Mother Earth, Gaia, I call you and all your high spirit guides of the elements. I call you at your place, which I appreciate and love. I thank you for all the beauty that you provide to us and for your generosity. In particular, I thank you for the daily food and drink that you give to me. This way I have a physical body out of matter that you lend to me. Without this I would only be a spirit. I thank you for having taken over the major and serious tasks from our Creator Spirit. How much you must suffer and endure because of humanity! We drill into you, and we make explosions in you. We poison and destroy your natural surface. I want to ask for forgiveness for what I myself have contributed to your suffering. I'm sorry. I know however, that you do not want to endure this forever. I understand that you wish to purify yourself, and shake off your ungrateful tormentors.

Now I would like to call the spirit guides of the elements in order to thank them as well.

I call to you, Great Spirit of air, wind, storms and clouds. I feel you every second, because you give me and all of us the pure oxygen that we need so badly. I thank you for allowing the birds, but also for allowing me to be in your element. I thank you for creating beautiful pictures with the clouds, and for distributing moisture on the earth so well. I love your mist on my skin. I thank you for the wonder of your ice and snow crystals, and the purity of the snow. I thank you beloved spirit of the air.

I call you, Great Spirit of water, springs, streams and rivers, but also the lakes and seas. I thank you for this precious and beautiful wetness that is life. I feel so connected with you, because three-quarters of me is made from you. Without you there would basically be no life here on earth. I thank you that you are home in your big waters, lakes and seas that you fill with so much life. I love you and I thank you.

Now I call you, Great Spirit of solid earth and the crystal worlds. I thank you for the wonderful and varied types and forms you show to us, that form the material upon which we live. I thank you for the grand mountains, *which both act as air-mass barriers and are the seat of your mystical crystal worlds. They hold your memory of ancient knowledge, but also your tears. I apologise for the over-exploitation, which occurs in and on you. I love you and thank you.*

I call you, Great Spirit of the mineral worlds. *You are the guardian of the minerals, oil, coal and many other treasures of the planet. I know that the Creator has wisely made provision for us and created the reserves for that purpose. And now you have to watch almost helplessly as the people burn those items to depletion. I thank you for the valuable products of the earth. I love you and thank you.*

I call you, Great Spirit of fire and light. *Without you, in the interior of the earth, our world would be made of ice. And with your innumerable angel helpers, you ensure the proper distribution of sunlight on Earth. You take care of every fire for heat and light. Without you, the planet would never have become habitable. I love you and I love the light in you. Many thanks.*

I call the Great Spirit guide of the collective consciousness of the animals on land and in the air. *You wonderful animals are our constant companions, and such graceful creatures that could only be created by God. You are so humble and full of love. You make so many sacrifices for us, without getting angry. I love you and thank you.*

Now I call the Great Spirit guide of the collective consciousness of the animals in the water especially the whales and dolphins. *You are the masters and rulers of the world's oceans, which are three times larger than the land portions of solid earth.*

I thank you for the richness of life in the oceans. I thank you that you tame the sea as your living area. I thank you for the great humility and love you have for people, and I apologise for all the people who persecute and kill ruthlessly. I thank you with love.

I call you, Great Spirit guide of the collective consciousness of the plants on land and in water. *You nourish us. I love your greatness. All plants are the living jewels of the earth. Without you the earth would be barren and empty. I love the flowers and their fragrances. I love and thank you.*

Now I call you sun, great co-creator spirit and brother, sister. *You are the greatest and most glorious orb in the sky. I thank you for your endless warmth and light. Without you, the Earth would be an ice star, and there would be no life.*

But now I call you, very greatest creator God *who has made all things with great wisdom, and infinite love and goodness to us. I thank you that you're not out of reach for me, but in my heart. Whenever I call you I sense your presence with me. I love you more than anything, because I can call you father and creator. I am a part of you, because you made me out of your love. I love you for what you are to me and give. Amen*

12.2 Prayer for Nature in the Spring

Oh dear God of Creation in me and in all, I call you to thank you. How wonderful and warm is your light on this Spring day, when all are waking who have been asleep for the Winter. Thank God I am healthy, and that

I can and may perceive much of your greatness. I am pleased with the first grasses and flowers, and with the first blossoms on the trees and bushes. After the long cold it seems to me like a rebirth.

The warm, life-giving light of our sun-body and co-creator from you, now brings everything to life. Because of this new wonder, my heart also experiences deep joy.

I smell the increasingly intense fragrances of the earth, the flowers and the blossoms. I perceive the smell of the water in the creeks and lakes, and feel the joy of the spirits of these elements. I hear the murmurs of nature, plants and earth spirits that are now really awake.

I thank you for the life of animals in the air, enjoying the warming light again, delighting, buzzing and rejoicing. I hear the laughter of children, now jumping around the houses. They also enjoy the spring.

Mother of light - Father of light, thank you for the new life of awakened nature. I am part of it, also my spirits awaken. Yes, the eternal spring and continuous feeling of happiness in my heart, only come to me through you. Amen

12.3 Prayer with and for the Sun

The sun – our planet of light and life. Without you there would be no Earth as it is. Without the light, that in turn transforms the nature spirits of the air element in warmth, there would be no life. Early cultures revered and worshiped the sun god / goddess. These people already knew, that without the light of the sun nothing grows, nothing lives. This is even noticed by us technological people, who are barely connected with Nature. After three weeks with a lot of rain, clouds and cold people and animals begin to suffer.

What could be more beautiful and more gripping than the morning sun rising behind the hills and mountains? When the darkness in which I still find myself is eaten up by the rapidly approaching pink morning light? The delicate morning light caresses the mountains, lies down carefully over fields and trees, and dries the dew covered grass. The grey fog floating over the ground gets an inner pink light, as if it is kissed by its lover, the emerging sun, and begins to blush.

When the yellow-red glistening ball of the sun first climbs upwards over the horizon, it seems as if the earth vibrates with the air layers. Over the slowly brightening ground, I see the spreading little particles move like waves to me. It often looks like an even surf on the sea. But the soft waves which gently touch and energise me are waves of glimmering light. This indescribable feeling cannot be compared to anything else. It is the contact of life-giving light with the spirit of my body and with my soul.

The wonderful sunsets are also magnificent and impressive. But it is the decay, the dying light. We know that the sun cannot go under and be extinguished, but this is how it appears. For our body this begins the recovery phase, because it needs rest, sleep and relaxation.

Our souls, however, are like the sun: constant, bright and active. They do not become tired just because a day has passed. This is easily observed with children who simply do not want to go to bed. The soul is indestructible and eternally alive. It is only temporarily connected with the body, a period of time together - and of course every new life as an incarnation. It is like sunrise in a new light. The soul is also hungry for light, yes, insatiable for it.

The darkness is almost unbearable to the soul, and causes suffering. But over the times and places man has lived, he has burdened himself with suffering and pain. Much has been done to him and more has been added from one to another. With every atrocity the soul is darkened, and increasingly separated from the inner light. Soon the man is so burdened that he can no longer enjoy his life. He can hardly bear the light anymore, and feels attracted to the darkness. Of this dark life there is more aplenty, and the supply doesn't stop. When the suffering person finally realises that his will is being manipulated and it is no longer himself that decides, it is already very late for his soul.

It is worthwhile to accept the light and as much as possible to capture it. I mean not only the sunlight but the spiritual light. The soul also needs a cleansing shower and a restorative balm, not just the physical body.

Prayer to the Sunrise

Mother-Father God of light, love and life, I call you! Thank you for the light of life that you give via the planet Sun. How wonderful is everything you have created, a good and beautiful life.
Now I call you, you co-creator spirit brother (sister) sun! How great you are, and how uniquely beautiful. The assignment you have received from the Creator is to emit light. Thank you for this life-creating light that radiates equally to all creatures on this earth. I am looking forward to this morning and to the whole day, seeing and feeling you. Amen

12.4 Prayer for All Creatures

This prayer may be recited as a meditation, very slowly, so that the participants can visualise everything word for word

Great God and Creator of all that is, I call you to say thank you for all the life around me. I am grateful for the work of your spirits from the mineral and plant kingdoms, those from the water and animal kingdoms, as well as the spirits of the air, fire and light realms.
Source of universal love, thank you for letting me be here now in the midst of your great plane of creation on Earth. Thank you that I am part of your whole, and that I can live as your child of God in this blessed time. I feel your light in me. This love that I perceive and live in is of you, a part of universal love. You give me the strength to complete everything with ease, hope and joy.

So I would like to invite you, as a perceptive, creative creature, into the warm Christ light at the centre of my heart. There, where your divine higher consciousness resides within me, this pink healing light shall spread itself and engulf me and my living room, the house, my village, the region around my home town, the whole continent, and soon the entire globe. Your red light of love illuminates and pulsates around the globe.

I would like for my love and gratitude to be felt and perceived by all the highest spirits of the elements, fire, light, water, minerals, and those of the rich plant and animal kingdoms.

Oh, beloved infinite Source, thank you for the joy and happiness that I already experience on this plane. My heart is blessed by the richness of your greatness in all that is. The small sadness that I feel about my physical transience is resolved by faith, my confidence and knowledge of the wonderful worlds of light that you have always been preparing for us. When I surround your living creation here with your light of love, my soul feels the longing to return back to you into the light. Thus I bring your light to our earth.

I pray for your healing light of love for all my brothers and sisters who are also with me on the path of knowledge. Lead us from our inner darkness to the light of Christ.

Creator God, thank you for my being, my senses and my feelings. With them, I can sense a part of the indescribable whole. I love you and your work. Amen

13. Harmonisation of Nature Spirits
(Dwarfs, Gnomes, Elves, Fairies)

Many people would scoff about this title, or laugh themselves silly, but they do exist, and many others too. These people, like the vast majority, are simply blind. But when they experience at some point that something is moving by itself, or is being destroyed again and again, they react with huge fear. Therefore these creatures belong with us on our planet. Perhaps we would not even exist without them? These various nature spirits are the links between animal and human souls. How animal souls are bound to their animal nature is addressed in the next chapter. Under certain conditions, however, an animal soul develops a powerful wish to be a being that has its own emotions and to live it's own life. This can happen, for example, when an animal has lived near humans for a long time, and his body, that is his flesh, has had frequent contact with the cells of the digestive system of humans. This animal soul may then be given an opportunity, as a soul outside of an animal body, to experience such an existence on probation so to speak. Depending on their disposition, this results in dwarves, elves and fairies. At first the dwarves look a little rough, but over time they develop traits that are ever more human. They are the link between the animal world and humanity. Mostly they live in the vicinity of people and their pets, and are quite shy. They can often be found around farms or houses, which are a bit isolated or have a garden. Such a place could be home to several generations of dwarves that would normally be no older than they were earlier as animals. They watch us, and want to learn from us. Elves and fairies are the links between people and plants, flowers, shrubs and trees. In winter, they also like to stay in the houses, and you see them as blue balls of light or little will-o-the-wisps floating around the room or among the houseplants. If we are content and happy, or especially if we pray, they become trusting and will be jumping around us. They love this kind of energy. When we are in a bad mood, none of them are nearby. If we speak to plants or flowers, we immediately attract many elves, which then give positive energy and good care to the plants. If we love to animals and caress them, there will often be many fairies or dwarves in the area, which are enjoying the love energy. If a person is evil and unjust to faithful, innocent animals, then the dwarves and fairies begin to fight for them. They can find this heartless and unfair. The elves feel or know when people are bad or negative to plants. The experiences with people accumulate, and are passed on from one to the next. It may be that suddenly nothing grows anymore on a plot or large meadow. Only in this kind of situation do people realise there is something going on beyond their perception.

'Nature spirits' are what I call the lowest level beings that are directly descended from earth material and the plant spirit world. They are usually composed of many spirits and adopt Gnomes as the strangest forms and modes of expression. Sometimes they look like tree trunks or potatoes, sometimes like mushrooms or small shrubs. They may well be very angry, when the land is cleared, excavated, tunnelled, or blown up, without first asking Mother Earth for permission and convincing the nature spirits of the necessity. Then they get very excited, and we can talk about luck when wind or storms come, and interfere with the works and carries them away. They combine with the air spirits, and in a flash can grow into a strong storm or hurricane. Such storms are unrelated to the normal meteorological phenomenon. These storms occur suddenly and without warning. But the real cause is the disturbance of the original structure of nature! Among these are virgin forest clearing, overexploitation of the earth, quarrying, explosions, and oil drilling. Yes, man is literally the world champion of the destruction of nature. But nature is stronger and hits back. It is not angry, no, it only creates the balance. We can never conquer nature or change it, because it just reacts to us. We are an important part of it. If the human species however is located at a higher level of consciousness than nature, then nature will obey us. We humans are called, with the power that resides and is alive within us, to intervene, to release and to liberate. At the place where you feel there is a disturbance, do the rite of clearance. Light one or more candles, and first, free yourself from earlier daily activities.

Prayer for Troubled Nature Spirits

Beloved Creator and Jesus Christ, in me and in all you have created, I call you here and now to help these suffering beings. Like us, you have given these beings the opportunity to create an identity for themselves on this planet and to develop themselves. I now call you, Angel of Light, Truth and Joy, to help me speak with these creatures, and help them:

Now I call you, countless nature spirits, dwarves, elves and fairies. I thank you for being ready to listen to me. Please come closer as I would like to get to know you better. Forgive me for not contacting you sooner. I did not know about you, your situation and your concerns. I apologise for everything I've done to you without knowing better. I also apologise to all my fellow human beings who know nothing of you, and as such have offended you. Forgive us.

I call on divine light of love from my heart to spread over you, and heal you (in deepest love, I visualise how red light of love from Christ flows from my heart, and spreads over the surroundings and over the spirits that are present.)

I ask you, dear beings, let go of your frustrations, pain and hate, and be free. Let go of your destructive rage, before it destroys you. For you are also subject to the same law of cause and effect that we are. Crave for God's light of love that dwells in all his creations, and also lives in us. Find it in yourself, and be healed and full of light. So be it. I thank you, God and Jesus, and you many angels for your love and help. Amen

Such a prayer can also be expressed in every gardener's plot or farmer's field. He will then have countless invisible helpers, to look after the animals and plants.

14. Animal Souls

In contrast to human beings, animals have directed souls. That is, they remain animals and do not develop further. After physical death, as a rule, they go back to the animal group soul, and come back again into this world as the same animal. However, they can evolve spiritually by proximity to people, and after a long time change into a higher animal species. Emotionally, an animal is built more simply than a human. But it does know joy and loyalty. It has likes and dislikes, and feels pain. It cannot understand 'why', but takes things as they are. But if an animal such as a pig or a cow feels at home in a place, it often returns after his physical death. But only if it has experienced fear or anger just before or during its killing. If one were to speak with the animal lovingly, and explain to it that its time is over, it would understand and voluntarily go to its death. This is because it has come to serve. But when this killing process is a painful and cruel one, an animal soul seems suddenly to lose its good nature. They themselves probably don't know how it happens. It seems to be their intention, at times like this to say "just get away from here, back to the stall, or back home". In this way, the simple animal soul loses its freedom, and by itself becomes earthbound. Now, imagine when such an animal soul comes back to the stable, it sees the other animals and they see it too! The people, however, think the dead animal is no longer there. The animal soul suffers as it no longer gets fed, and feels completely ignored.

I don't know how much it suffers or feels pain, but the potential for negative energy certainly increases. This I always see and feel. The other animals in the stable also suffer, because they are afraid of the ghost that is visible to them. It is actually the same as with us humans! Take, for example, a father, who as long as he lived was the darling of the family. If he comes back as a spirit after an accidental death, all of a sudden his family feel fear about his death. But he is, and still feels, as he always was. This is very sad, because humans seem to be wearing blinders, or are just totally unaware.

Animal souls do not necessarily have to stay with their companions, and can often be found in empty barns. They also often seek out sensitive people in their homes. In such places it is not possible to get a good night's sleep.

14.1 Prayer of Release for Animal Souls

Dear Creator in me and all that is, and you beloved Jesus Christ, assist me with all the angels of light, truth and joy. Help the poor souls of these earth-bound animals, to release themselves. Translate my words, so that they can understand me.

Now I call you, countless animals, who are here and feeling unhappy. You are dead animals who have not noticed or felt your own death. You had great fear of death, because you were not prepared.

You could not understand what happened to you, and you have now taken refuge there, where you came

from: in your stall, on your pasture, or even in the house where you have lived together with humans.
I would like to apologise for the people that have tormented and despised you. You are wonderful animal creatures, made by the hands of the Creator. Now let go of your fears, your frustrations and your anger, because you cannot be free and go away from here, if you do not let go of your old life and home.
I ask you, dear angel, help these animal souls find the way to their destination. In the name of our Creator, Love and Jesus Christ, go now and be free. So be it. Thank you. Amen

If the interference with you or your children persists, then repeat this prayer in the moment when you feel disturbed, if necessary even during night-time hours.

14.2 Prayer to the Animal Souls

When you take the time to look at the animals as they live in your area, you realise that they behave mostly out of kindness and humility. They are there and they serve, and seem to expect nothing. You never really know what they perceive and think about. And yet they have feelings of affection and loyalty above all. I mean, of course, our pets and farm animals, but also the birds that fly around the house. They perceive us and feel how it is going for us or how we feel. Clairvoyants can even see a beautiful aura of the reddish light of love around wild animals in the forest. We can address all the animals and send them our love. The love of the animal creatures is returned to us, somehow, sometime. When someone tortures animals and despises them, the effect on humans is negative karma. And this is suffering energy that remains bound to people, which doesn't go away by itself or simply disappear. Such Elementals of suffering attach themselves to the perpetrator as compensation. No one can possibly believe that if I torture the neighbour's dog or cat, it doesn't matter and will not become a burden. No, the law of cause and effect always comes back, and never forgets anything.

Is there anything more beautiful than animals in the pasture or stable coming to receive a loving prayer? But this is also good to do for the animals at the lake, in the woods or in your garden. You will see that this prayer is most effective when you speak out loud and ask the angels of love to include the animals that are not directly present.

Prayer

Beloved animal souls, I call you. Please listen to me:
I am happy to live together with you on this planet. You are wonderful creatures and decorations for the earth. You love people and serve us, although many of us do not notice or appreciate you enough. For this I apologise. I love you and your souls. I surround you with the pink light of love and the warmth of my heart.
Thank You Mother Father God for all the wonderful, great life in your animal kingdoms in both the micro and macrocosms. Thanks for the abundant bird life that en-

riches our lives, and thank you for all the animals that live with us. I want to send my love to all animals under the ground and in the water. Thanks for your universal love and divine healing power, in which my loved ones and I are embedded. Amen

15. Purification from the Lowest Earth Energies

As covered in some detail in this book, such lowest earth-material energies may only be washed away and cleared with 'Blessed Oil' (light oil). This particular form of energy is still on the earth-material vibration level. Although recognised by spiritual and clairvoyant people, it does not yet have any consciousness. It cannot be addressed or cleared by prayer, for it doesn't know it exists. Even regular (unblessed) oil or multiple showers will not help when one is attacked by it.

Oil Blessing
Suitable for this purpose is any edible oil, in particular because it is safe for skin and hair, an olive or sunflower oil. Place the oil on the table and around it a couple of burning candles. Speak aloud the prayer:

Dear Creator God and dear Jesus Christ. Please bless this oil, the product of your soil. Make this oil as bright and shining in all colours, that it undermined the low earth energy and can become detached from any human being. I thank you, God and Jesus, for your help and love. Amen

Procedure for Cleaning

The person that you want rid of it is put at the top in the bath (or outside in the garden) on an old cloth or paper towel. First, you rub your hands and arms up to the elbows with this oil, so you will not become contaminated. Now the afflicted person should rub the oil richly onto the scalp, behind the ears and the neck. Ask that the angels should help clean the body at the same time. Rub the back of the person's body, always from top to bottom. The afflicted should further rub the front and all the other parts of the body, including between the legs, the genitals and the soles of the feet. **Ask now with a loud voice, the earth mother Gaia, that she should take back these low earth material/energies. Thank her and the angels for their help.** Then pat the excess oil lightly with a household cloth. At all costs, leave the oil overnight on the skin and only take a shower the next morning.

This low earth energy looks like ice crystals that grow on a cold window glass produced as if by magic. Only the colour is not white, but cream-to ochre or brown to cacao coloured. It does not look at all pretty or pleasant.
I don't recommend storing any waste oil in the home! All used towels or paper products should go into the garbage or be burned. The previously afflicted person realises that he/she is free, if the previously affected areas of skin become rosy (immediately after cleaning) for a few hours, clearly different from the previously non-affected body parts. The person might feel a later

relapse by local burning or biting. In such a case, a repetition of the rites would be necessary.

Should the person have pets, they should keep a certain distance from it. Often the animals are indeed themselves the carrier or bearer of these lowest earth energies. But I do not yet know whether or not they suffer from it as people do. There is an easy way to rid the animals of it: You take an empty household spray bottle and fill it one third with blessed oil and two thirds with spirits or liquor. This way we can spray the animals from a distance from time to time. They do not like this but it helps.

16. Wealth and Poverty

Many people are poor on this earth. Poor in goods and food, but also in spirit. At some point someone told them that they could only reach the kingdom of heaven if they renounce all riches here on earth. This is true to some extent. We cannot take our wealth with us, for it would limit us in the spiritual world and only constrict our development. But, who can testify that God ever wanted us to make a vow of poverty? Would we as parents have any joy, if our children are in rags and die of hunger? Certainly not! We are God's children. Let us consider nature. Every single plant produces abundant flowers and thousands of seeds. This is truly an incredible abundance. Also, an incredible variety of animals exist and evolve in abundance. But there are places on earth where nothing grows or can grow. These are areas that previously flowed with milk and honey. People have gone there but turned away from God, so he took away their livelihood. Now people in later incarnations have to live there and learn humility. This is the law of cause and effect. Only when these souls return to the true light in themselves, does the bounty of nature return relatively quickly.

We should think about whether we have not overused our fertile regions, and as a result are punishing ourselves with the same fate. Poverty and shortages are, in any case, being revisited in current consciousness. When someone envies the richest of the rich, resentment and even hatred come back in return many times over. This is a basic law of the spiritual worlds and our duality: **If I begrudge another, I prevent myself from having what they have too!** And because I do not realise it, I will get poorer and more dispossessed over time. I can change that, however, and quickly. I have to stop my restrictive thought patterns, and align them with the universe. However, if I beg, the creator will not recognise me. This is because God has not created beggars. **We are his children and have a fundamental right to wealth and abundance.** But we only get it through him. And only if he feels that we love him, because only then we are in contact with him.

16.1 Suffering Due To Obesity

Many people suffer from chronic obesity. They buy too much food, and continuously overfill their fridge. They feel forced or pressured to do so. Later, they have to throw much of it away. Their enduring hunger cannot quite be satisfied. Diets are useless because of the well-known yo-yo effect, the weight comes back quickly. Many have a gastric band fitted, or even remove part of their stomach, but gain weight despite of this. I found that the cause of the problem is in their **soul's past**. Countless people have ended previous lives in a famine. Some have even done so several times in different lives. A few of them have the problem of anorexia, they cannot increase their weight.

When such an event takes place, a great suffering occurs at all levels if consciousness. The mind, in its upper and middle levels, registers awareness of the problem and stores it. As a result many different feelings of suffering will be endured. "I see all my loved ones die and know that I will die in the coming days."

At the same time the body spirits register that they are not getting enough, feel hunger and suffer as a result. All this is recorded and remains stored in levels of subconscious. The only solution is to delete the experience and resulting commands at all levels. Please be patient and fight for yourself, you're worth it!

Ritual of Liberation from Obesity

Dear God and your divine angel of healing, please help me: Beloved Great Spirit of my body, I call you. I am familiar with you, since I have a body that is firmly bound to me. But because of the ignorance of my mind, I have more hindered than helped your important work. In a past life I died of hunger. I suffered from incredible fears before that death, and they are stored within my emotions. In this life, I do not need or want them anymore.

• I call to the consciousness of my body spirits that became ill from the distress of my death • I call on all the patterns of suffering caused by starvation • I call on all patterns that have been collected, recorded and stored • I call on all the patterns that want to keep eating more and more and never be satisfied. ***I do not want you anymore, I free you from me.***

My dear body, my body spirits, I want to apologise to you for the negative commands I have given you in this life, but also in many other lives before. I would ask you that all misery of death, fear, distress, pain, injury, disappointment and anger are released and discarded. My dear body now erase everything that is against your actual wholeness and good health. That is what I would like. Remember your fundamental purpose, your basic task, namely, wellbeing. My dear body, you're perfectly whole and healthy. You are healthy, slim and beautiful. I love you. I thank you for your great and important task. I love you and I apologise again.

I beg you, dear great Creator, help my body with your angels of healing. So be it. I am healed. I thank you and love you because you have created perfection in me. My thanks and my love goes to you, Ultimate Creator and to the ever-present light of your love. Amen

16.2 The New Era

During this time of change to the new energy of the earth, there will also be a major change in the economy. Supposed values disappear and make room for spiritual values. In the foreseeable future, there will be days when the value of the euro and the dollar decline by 20-30%, until these supposed currencies disappear. But also the apparently rock-hard Swiss Franc will become weak and finally disappear, because it has become a world-wide prostitute. For people who only believed in the power of money, it will be extremely difficult to find their way around in the new situation. For each one of us is facing a major challenge, when even a lot of money and gold is not enough to buy bread any longer. The opinion that I only get something if I give is outdated and will not quite function anymore, when I no longer have a medium of exchange, or more valuable asset, i.e. money. The thinking will have to be: **I give of my heart and I'm happy if others have a lot.** When everyone starts to think this way, everything will come back in full swing.

Mother Earth is changing together with the nature and frequency of its vibration. Hence the man who stands at war with nature, so to speak, will immediately notice that it no longer feeds him. More precisely, she is no longer prepared to do so. Without respect and love for her, but also for the animals it will come about that people will starve with full plates. The new consciousness will be: **I am a part of the whole. Nature, flora and fauna are the basis of everything. I respect and honour them. I love them as much as I love myself.**

Prayer for Abundance

Beloved Creator God in me and in all that is. I call you and all the angels of love, truth and joy. Help me to free myself from old behaviour patterns of scarcity.

I now call my old Elementals and behaviour patterns from this and all previous lives and realities that produce poverty and scarcity. I have created you without understanding the consequences, and now liberate you from these tasks. You are free, and I am free. Beloved Creator and dear angels, I thank you for the wealth and abundance that surrounds me, and is now revealed to me. Thank you for helping me to live in harmony with nature and the animals. Thank you that I can live well without the corrupting power of money, and can find my way.

Thank you, Father-Mother-God, for the gifts to your children. I thank you with all my heart. I love you and your angelic light beings. Thank you. Amen

17. Wishes and Requests to Ones Own Higher Consciousness and to the Divine Spirit Worlds

"Everyone makes wishes, but only God knows if they will come true." So goes the old saying goes. But very few people know that they themselves stand in the way of their prosperity, and the fulfilment of their wishes. Why is that?

Can a wish manifest, if I have thought or requested exactly the opposite in this or previous lives? That is about the same as when two powerful draft horses pull a heavy log of wood, each in opposite directions. Since they are only toiling in a tug of war, they suffer for nothing.

Also in our lives, there is only forward, when we realise that we were and still are pulling ourselves backward. So it is a matter of letting go of, and liberating us from old requests and wishes, which have been or are still directed against us. For example: I'll never be happy • I am not allowed to be happy • I was born poor, and will die poor • I'm ugly and never find a partner • Everything belongs to others, and I will never have anything • I must suffer, because the church tells me I must, in order to gain the kingdom of heaven and paradise…

These are all commands and manifesting wishes with which we are living in the here and now. First I have to get rid of all things that are old, which have consequences I am not conscious of, and which continue to give me new feelings of guilt. I must completely forgive myself, with love.

17.1 Wishes and Requests

Example 1: A New Job

For anything new that is wished for or requested, it is essential that it is well thought out and does not harm anyone. Let's take for example a new job or new career.

So, I want a job, but I won't get it by begging for it. No, I have to manifest it: I reflect on it as if I already have it, and as if it already gives me joy. The preparation is the most important thing, and creates a good, new job.

Procedure

I imagine a white stone obelisk, with four sides free for writing on, and a pointed top. Each side is covered in writing, and it all works together as a whole.

First Face:
Who I am, and what I would like to manifest belongs here:
- Name, birthday and address
- I am the I Am in the here and now
- I wish (request, bring in) all the things that I am going to write on this obelisk

Second Face:
I imagine the new workplace:
- The work location is easily accessible for me
- It has a bright and calm atmosphere
- The men and women are nice to work with
- There are no disease-causing fields or microwave radiation (cell phone antennas) nearby
- The work gives me joy
- It guarantees me a good, regular income

Third Face:
This is what I want to give:

- I want to work hard and use all my knowledge and ability
- I will use discretion and be faithful to the company
- I will work together with my colleagues as a team

Fourth Face:

This is what we will achieve together:
- I would like to integrate optimally with the firm
- I want to involve myself, as though the company belongs to me, but always for the benefit of all
- Together we are strong, which makes me happy as it does all the others

When all four sides are well considered and well described, I send out the request with approximately the following words:

Prayer

Beloved Creator, Father-Mother-God, I hereby submit this wish to you, and request that it be manifested in the here and now.

Thank you for all that I have and for all that you give to me so easily every day. I thank you and your beloved angels for all your assistance and support. Thank you for giving me this job. I love you. Amen

Example 2: Desired Partner

When making a wish for a desirable man or woman partner, one likewise imagines a shiny white stone obelisk with writing on four sides.

First Face:
Who I am, and my intention to manifest:
- Name, birthday and address
- I am the I Am in the here and now
- I wish (request, bring in) everything I have written on this obelisk

Second Face:
Now imagine your ideal partner specifically. Paint yourself a detailed picture if how he/she will be, and even how he/she should not be.

Here are some examples:
- They shall be kind
- They shall be intelligent
- They shall be faithful
- They should be hard working and reliable
- They should have regular work and a good income
- A family man that loves children (provided that children are desired)
- He/she must absolutely be available for a new relationship, and free from prior commitments
- He/she should be clean and smell nice
- Perhaps he/she should have a house

Don't forget to include: The attributes that he/she should not have (such as not fat, not bald, not too much hair, if you do not like these, or non-smoker, or they should be free of addictions).

Let God know that you do not want a mother/father or a sibling from past lives. You no longer want a former husband/wife. You want to experience new things, and not be spiritually stopped by repeating old patterns.

Think about exactly what you want to include and exclude. The more details you create on the list, the more accurate is the 'delivery'!

You should not, however, go into all the details of outward appearance. There is, of course, a higher purpose behind this. What if the spirit world has found exactly the right partner for you, but could not bring them to you, because they unfortunately do not have the correct eye colour, and otherwise in all respects would fit wonderfully? Do not limit the angels with trivia! Ultimately it comes down to affection at the heart level.

Third Face:

Here confirm what you can and want to give!
- A humorous partner
- Loving and tender
- A good parent
- Faithful
- Maybe you want to remain self-employed and independent
- Or manage the household with joy
- You bring a good deal, for example you may speak several languages
- Or you have musical talent
- Etc.

Fourth Face:
Now you confirm what you will do when you are connected with one another:
What do you want your life together to be like
What do you want to undertake when having a life together?
(What good is the most handsome man, if he is hanging out with his friends every night or goes to sports events every weekend in which you do not have any interest? What if he turns away from his own problems or would rather be a sport's addict?)

Suggestions: My wish is…
- We spend much time together
- We raise the children together
- We go walking together
- We work together in the garden
- Full of joy, we design or create full an apartment or a home together
- We go dancing together
- We go on ski holidays together
- Joint cinema, concert, theatre evenings
- We gladly invite friends to our home
- We both enjoy family gatherings
- We travel the world together
- We have common interests and understand each other completely.

Think carefully about everything you want to do together, because you get what you ask for (and then you have it!)

When you have completely filled the obelisk with your ideas, then it's time for the wishes. All wishes are placed as a package with the Creator, to whom nothing is impossible, when you call him as follows:

Prayer

Dearest Creator God in me and in all that is, I call to you. Thank you for everything you give me and for the requests you have already granted. I would hereby like to give you my request for a life partner, with full confidence that nothing is impossible for you. I know that you love me, and that you are always there. You feel joy when, as your child, I am happy. Thank you for looking out for me, and for reserving this partner for me. I love you and your Angel of Joy. Amen

17.2 The Joy

Should you, dear reader, find yourself in the lucky situation of not spending your life alone but with a lovely partner, you should be thankful to god each and every day and cherish this by expressing your joy to your partner.

For a Man – Prayer for His Beloved

Beloved God and my Creator, I thank you for the love that flows from my dear partner to me. Thank you for the joy and happiness that I feel as a result. In my heart I feel love and warmth for all that lives around me. Thank you for the gift of being able to feel and receive love, but also for the ability to pass it on. I feel your great love around me when she caresses and touches me. I stroke her skin, and thus her soul, in which you live. Your universal love has been there since the beginning of her life and always will be. Give her the gift of being able to feel and perceive you. This would increase my happiness and even strengthen my love for her. Beloved creator of her and of all, I want all people to feel this way and to be as happy as I have been in this body in this world in the now. Thanks for making me feel joy and for being able to see and perceive everything in the best light. Your light shall grow in my heart above all, and my longing for you and the light-filled planes shall diminish, so I may gain more experience. But I know I can go back there when I have removed all my unhealthy patterns and vices. How nice it is to be in the world. Amen

For a Woman – Prayer for Her Beloved

Beloved God and my Creator, I thank you for the love that flows from my dear partner to me. Thank you for the joy and happiness that I feel as a result. In my heart I feel love and warmth for all that lives around me. Thank you for the gift of being able to feel and receive love, but also for the ability to pass it on. I feel your great love around me when he caresses and touches me. I stroke his skin, and thus his soul, in which you live. Your universal love has been there since the beginning of his life and always will be. Give him the gift of being able to feel and perceive you. This would increase my happiness and even strengthen my love for him. Beloved creator of him and of all, I want all people to feel this way and to be as happy as I have been in this body in this world in the now. Thanks for making me feel joy and for being able to see and perceive everything in the best light. Your light shall grow in my heart above all, and my longing for you and the light-filled planes shall diminish, so I may gain more experience. But I know I can go back there when I have removed all my unhealthy patterns and vices. How nice it is to be in the world. Amen

18. Suffering and Pain of the Body

Our body is a marvel of nature, as they say, but also an even greater mystery. To clarify: as a soul and a spirit, so with the mind, we are not the body. It is an autonomous, more animal-like structure. It runs like it is self-controlled, and if we do not disturb or interfere with it, it works perfectly. When it is born of a mother, it is still so small, yet it can do almost everything except for motor skills. This is true, even though the mind and perceptions are still asleep. This makes clear and proves it is not us that control our body by means of consciousness. We are also not in a position to control consciously the heart rate or cell division, let alone our growth. If it shows disorders, pain or chronic suffering, that even doctors cannot explain, it is either injured or has been programmed incorrectly. Neither begging nor prayer changes anything. It just stays the same or becomes worse.

As a parable, one could imagine the following example. A woman gets an old dog that belonged to a man who has died. She has never before seen the man with his dog, and knows nothing of the animal, such as how he was brought up. Now the woman wants him to behave like a lap dog, sitting on her lap and going to bed with her, etc. He reacts to these invitations by wagging his tail, but does not jump on her lap, let alone go on her bed. When she picks him up and forces him to do so, the dog begins to whine, pulls in his tail and jumps as quickly as possible back to the floor. He cannot because he was trained that this is an offence, and was punished immediately. He is now programmed.

Let's look at our body more closely. It is, like all seemingly solid matter, pure spirit and vibrates with its small particles in every cell. Each organ is arranged in a hierarchical structure, and organised accordingly. Within each organ are countless little nature spirits with one boss. They are directed and have been given their permanent job, by the creator of the highest wisdom. These organ chiefs are further joined together in groups to carry out higher-level tasks or take commands from our spirit, even sometimes imprudent ones.

We say commands often without much thought: "I don't want to see that. I can't stomach watching this anymore. I don't want to hear that anymore. I will not take notice anymore." It is also the case with "I do not want to smell anymore". These are all commands to our body. Fortunately, it does not always obey, as previously mentioned for the old dog. But one day the body responds and makes it stop after all. You will hear people say, "Oh, my eyes have gotten almost blind in one week" or "Suddenly, I can't feel or smell anything!" In these cases, good advice is really expensive, or completely unobtainable.

Think about what happens during a heart or organ removal. The organs continue to work as if by a miracle. They were programmed to work, and even under the most adverse circumstances, continue to function totally autonomously. These removed hearts, livers, kidneys, etc. twitch and still live up to 24 hours outside the body, provided that the highest chief of the body, the soul, has not given the command to shut down. Otherwise, the programmed life-spirit just wants to live and serve. The idea of it, and also the execution, comes from the highest spirit of life and love. When disturbances in the system, or pain, are exhibited, it is man and his stupidity or bad events from past lives that are to blame. Some people think that if their father had problems with the hip joints and knees, then they get it. Others say if their mother and aunt had problems with their breasts and died, then they are also affected. Or think of others that have to die early, because their family have all died young for generations. And how many people believe that they are at the age of all sorts of ailments such as arthritis, Alzheimer's or Parkinson's disease? If these 'orders' are often given out, and it is firmly believed, then they will surely arrive.

Our body is actually designed for prolonged use. Humanity has learned this as well, because we live longer now than we did a generation or two ago. It is believed that medicine helps to increase life expectancy. Yes many people vegetate with great suffering, and the lowest quality of life, which has nothing to do with a good or a normal life.

I have had success countless times in many people with the following prayer and ritual of liberation. Try it, but with your firm conviction and with your entire will. **Do it earnestly with love for your own body!**

(For further benefit, I recommend you set the mood by first doing the short meditation "Clearance and Recovery from Patterns of Suffering in Your Own Body", Chapter 9.2)

18.1 Deprogramming Body Suffering, Sickness and Pain

Beloved, great spirit of my body, I call you. Also in particular, I call you, guiding spirits of my eyes, my optic nerve and my brain stem. (You can contact the guiding spirit of any organ.) *I am aware of you, since I have a solid body. Because of my lack of knowledge and lack of understanding, I have hindered more than helped your important work. I would like to apologise for the negative commands that I have given you in this life, but also in the many lives before. I also ask you to release and discard all illness and injury, as well as shameful torture, that I suffered in the past.*

Now release everything that is against being healthy, and against your true wholeness. That is what I would like. Remember your basic purpose, your basic task, namely the soundness of all body functions. You are whole and complete. I thank you for your great and important work. I love you and I apologise again. I ask you, beloved creator and master builder of my body, help it with your angels of healing. So be it. I'm healed! I thank you and love you, since you created perfection in me and have given it to me. My thanks and my love is for you, Creator and the great light of your ever-present love. Amen

This prayer may be used for all organs. Particularly, it is essential for patients who carry a transplanted organ in their body. In this case, a welcome and a thanksgiving ritual is called for and is very important. Conscious acceptance and gratitude, regularly enveloped in love light, helps the new organ to settle in and do its work.

18.2 Relief for Allergies from Pollen, Trees, Shrubs, Flowers and Grasses

Unfortunately, allergy sufferers become more and more ingrained with the idea that they really are suffering and that total relief is impossible. Many believe more in chemicals and destructive cortisone, than in a Creator Spirit and in our own healing power.

All people are being informed on radio and TV, but also in every newspaper, now that the pollen count of this shrub is up and then again that another is rising. These messages are in themselves completely inane and take advantage of people. Most become confused and their suffering only increases.

A standard prayer will not help because we ourselves have created these patterns of suffering. Everyone wants a true improvement in their quality of life and not the opposite. What do we do?

First, we must recognise the old patterns of suffering, call to them and then release them. Perhaps we have already been carrying these around for many incarnations. Each time they become stronger. After the liberation and clearance of these patterns of fear (elementals), we can re-program the body's spirit.

Prayer for the Relief from Allergies

I call you, my Creator, with your servants and angels. Dear God and angels of truth, of light and of pure love, I ask for assistance and healing. Please help me to find and clear old patterns of suffering in my body and in the various levels of my consciousness.

I now call all old experiences and self-created afflictions from previous times that I have accumulated on all levels. In particular, I call to those that cause me to have fear for nature and its flowers and pollen, as well as for insects, such as mites, spiders and ticks.

I now live in this body in this world and in this time, and I no longer want you. You and your patterns are free, and so am I. Beloved Angel, please assist me with this. Now I call the master consciousness of my body. Your controlling spirits represent my body. You were disturbed and besieged with false commands. Please now delete all incorrect commands that were telepathically programmed by me and by other people.

Above all, I speak to you, controlling spirits of my sense organs, eyes, nose and ears. You remember how you were created at the beginning of all being from the Creator. You are now totally and completely healed. You are very important to me. Together with me you will be healthy until the last hour. Now delete everything that works against this. I love you and I love my body.

I thank you God that you have created me whole and healthy and that you help me to be free from my own judgments and fears.

Thank you for all the beauty in your wonderful nature. My body is a part of it. I thank you for the beautiful shrubs, trees, flowers and grasses and for the insects. You are like me, a part of the whole. I love you because you are the jewellery and decorations of our planet Earth. I embrace and love you. You do me good.

Thank you Father-Mother God for your universal love in the greatest as well as in the smallest. Thank you for letting me be here in the variety of your unfathomable creativity.

Give me the power to love the smallest with the whole expression of my heart. This gives me the hope and joy of discovering your greatest and most glorious creations, and of integrating your universal love in me. I thank you for your light that shines in me and shows me the way to you. As your child of God, I say thank you to your angels of love. Amen

19. Casualties of War and Release for Suffering Victims

Many people in this life are victims of war. They and their children are often deeply traumatised. I know of school teachers who had such children in their class. These children would suddenly jump under their school desks in fear, when they heard thunder or a supersonic boom. They would tremble all over for almost an hour. Think about what they did to the people in the Balkans, or the Kurds in eastern Turkey, or the Armenians and Palestinians. This hateful ethnic cleansing, and also the expulsion of entire populations, has a powerful effect on later generations. An enormous potential for violence is created by the people's collective karma. These accumulated energies of hatred that originate from suffering, can also be suddenly discharged several generations later. Many of these people live as refugees with us or in other European countries, and often had big problems with everyday life as children. When I ask them why they cannot sleep they almost all say that they still have war nightmares. Unfortunately, as parents they often hand down to their children the same fears and panic attacks even when they are born much later. When these children ask where was this or that uncle or Grandmother, they get the answer, for example, that they were kidnapped and killed by Serbs, or that the Turks had murdered them by the thousands. This makes an impression on them. The fears that arise from this cannot be explained away or simply ignored, because they were experienced and are real. The truth may even be exaggerated and presented from a one-sided point of view. But there are no coincidences in life. Everything that happens is the result of the law of cause and effect. In order to better endure the effect of all these atrocities, it is greatly beneficial for them to increase their awareness and break the spiral of suffering.

Prayer for the Release of Wartime Experiences

I call you, dear Creator, and all your angels of love, truth and joy, for assistance. I would now like to call back all my sad experiences, pains and fears once again into my conscious daytime memory. I would also like to retrieve all the painful old burdens from previous lives that I am no longer conscious of, so that they may show themselves to me. I know that these actually all belong to the past, but they continue to come to resurface again and again.

Now I ask you, my creator, release me from these old sufferings, pains and trauma. I hand over to you and at the same time forgive all former enemies and opponents from deep within my heart. At the same time I do not know if I have a clean slate, if I have tortured, persecuted and killed others. From my heart, I ask all these potential victims for pardon and forgiveness. Beloved angels, please help ensure that these thoughts are received by all concerned. I ask you, great Creator, to heal the old wounds of my victims and mine too. Thank you for your love and the light that you give to me. Help me to learn my lessons from this, and protect me from the temptation to dredge up old patterns again. I thank you and all the angels of love, joy and truth for your guidance. I am free and I thank you, oh God. Amen

20. Birth and Death

20.1 Baptism Ritual

For many people, baptism is a mysterious procedure that is full of riddles. Yet Jesus was baptised by John the Baptist in Jordan. One wonders why the highest spirit considers a baptism in a river to be in the best interests of a man, and what it does for him?

For the man Jesus, who obeyed all the spirits of nature in all its elements, baptism had a kind of connection character. He wanted to show people I am one of you, no one is more than you, just as no one is less than you.

The baptism of Jesus can be understood as an initialisation to and connection with the divine. Whereas parents and family members will enjoy this baptism more when done for an infant, and thereby initiate a Christian path for the child.

In Jesus' time only adults could be baptised, to accept the Christian faith and to let go of their old one. People in the Christian churches were instilled with the fear that those who die without being baptised go to hell. This is of course a first-class intimidation tactic and even brainwashing, which causes fear of death. "Oh, my child will go to hell, if it has an accident and dies, and I alone am to blame." But the goal was only to bind their believers and offspring to the church. You can well imagine that they could not take responsibility for their young children nor for the grown ones. The child has an old, adult soul, and is possibly more experienced and more mature than you. Therefore it is perhaps more important to receive the real baptism as an adult, so to speak as a confirmation or love contract with the Supreme Creator Spirit. But for parents it can be reassuring when they have named their child and know the covenant with the God of Love is signed for their child. It is also meant as a gift for the new life that has come to them.

Baptism Ritual and Prayer

(It is irrelevant whether it is the parents or a friend that say the prayer in the presence of the infant.)

We call you beloved Father-Mother God with you Jesus Christ to us in our midst. We thank you for this child who has come to us. It is an old soul who wants to experience a path through time and space to gain new knowledge or to remove old guilt.

We welcome you here in this family. There are many things we can see through you and re-learn. We mean well, we may teach you something new, but as an old soul you know everything already. So we particularly want to give you love and security. We want to be there for you, to help you to achieve your goal in life and fulfil your life plan.

Dearest Saviour Jesus Christ with all your angels of love, joy and the power of love, bless this child, his body, his spirit and his soul. Help him so that he can go through life easily, and let him always be aware of his spiritual freedom.

Dear Saviour, bless this water with your highest and purest light, your love, and the oil that comes from the spirit of the earth and the plant.

(You take a little water with olive oil and wet his forehead and the surface of his scalp, the crown chakra.)

In the name of our God of love, we baptise you with the blessed water of eternal life in the name …(name of the child).

May the Holy Spirit come to be associated with you always to guide and protect you.

Beloved …(name of the child), *the blessed olive oil from an ancient tree, sacred as the holy goodness of mother earth, would like to cleanse your chakra and keep it open. Let the light of your soul shine together with the light of Christ in your heart. May your path may be full of light and joy.*

Beloved Jesus, bless the parents with your love so that together with …(name of the child) *they may experience the joys of heaven with your gift of light.*

Great Creator, we thank you for the health, wealth, and for the abundance of body, mind and soul. Bless all those present with your love and joy.

We thank you. Amen

20.2 Dying Process, Funeral Prayer and Burial Rite

For many people who have died or have just passed over, a clearance is extremely important. It is precisely at this stage that the soul is usually still undecided or very confused by what just happened. The worst that can happen to them, is that you do not let them go away. An example is when a survivor cries and screams, "You can not walk away, you can not leave me alone, you must not go." That is good news to the dying person. It feels like a man hanging on a tuft of grass on the edge of an overhanging cliff. He can't see the others, standing watch over him, but can hear them clearly. He senses that something is happening to him, however he cannot change anything. The others above cannot see him either and yell that he should not let go. He must have a feeling of powerlessness. In this unfolding tragedy, those of us remaining are stronger than the one who just died. This means that our mental powers can bind the newly dead to us. There is, therefore, an urgent need for us to change our death cult. We would facilitate the enlightenment of countless souls and ultimately our own as well. The following funeral prayer can lead the way to this end:

Death-Release Prayer

Dear Creator God and Saviour Jesus Christ, I call you with all your angels for assistance. Help …(name) to hear me and enable him to wake up at this time. I call you …(name), wake up, because you have passed away and left your body.

I would like to assist you, so you can be free. You are now free as a soul and spirit, and you are outside of your body. Please also let go yourself of your body and its awareness of the material world. Become a free and pure spirit. Don't let yourself be bound and held back from the important tasks and work you left behind. Even if the old matters are not yet complete, they are not important anymore. They can wait or be taken care of by others on your behalf.

Make yourself as free as a bird or a butterfly in the air. They do not worry about tomorrow, because they are just in the now. And so are you. Now you're going on a holiday to the light. You have done your work here and it is good, very good, believe me. You're going back to your origins, back to the light and to joy. There at home again, you will be newly re-born. You return home in your light, to joy and your happiness. Although we are all sad, we know that your free will is what is important for you now. We acknowledge this with pure love. For this reason, we want to let go of any sadness. We want you to see that we are filled with great joy that you have arrived in your paradise of light!

We therefore ask you to go now and leave this house and this area, so you do not lose your way there, or find yourself standing in front of closed doors. We encourage you with all our love and prayers, to be on your way knowing that you go with our blessings and support.

You know that in our hearts we are all always connected. Nevertheless, a divine unconditional love is freedom. We allow this for you, as you do for us. So now let go. Call up the divine light within you, and become free from all that still holds you back. So be it.

Dear Saviour Jesus Christ envelop this house with the purest pink light of your love. The red-coloured light of love that shines from your heart refreshes and strengthens …(name) and us as well. Then we can feel your universal love at any time. Thank you for the wonderful opportunities that …(name) has given us, namely, meetings with kind and loving people. I/we thank God for this life and the lessons of previous life. I thank you that we are a part of the whole and yet we still may exercise our free will. I thank you for all the beauty that surrounds us now and forever. I thank you for all the opportunities that you give us to recognise that you live within each of us. Look at …(name) and help him/her with your immense love, which is also a comfort for us. The light of your

love is always around us and in us, and protects us from the darkness. Guide us safely when we lose our way and give us your strength and confidence. When we feel your love and warmth in our hearts, we know you are there and nothing can hinder us. We are your children and live happily forever with you and within you. So be it.
I love you, Almighty God and Jesus Christ and all your supporting angels of pure love. You are always welcome in me and in my heart. Amen

Many of the dead stay with and around their relatives for a long time because they have not been released. Also, the deceased may be present at their own funerals, but they are very often half asleep or in a coma-like state. Therefore, it is to their benefit when they are called and awakened. With their newfound alertness, they can move and develop much more freely. As the funeral for the earth bound soul is rarely beneficial, we can give them support. Very often, the deceased went with an internal grievance or under distress from us, and therefore don't believe anything any more. They think they are lost for all time or surely damned, perhaps for all eternity. They have in the meantime, mostly realised how many unresolved problems or tasks are still pending or are yet to be resolved. This makes the deceased even more dejected and depressed. The total sense of being knocked out applies here. The desire to turn off or release is getting stronger. For most, one can describe the feeling as a so-called 'Death Depression'. This is only true for the earth bound souls, though these make up the largest share. Those souls, who enter into the astral plane and are completely disembodied, are taught by high spirits and comforted. What a huge difference for those who remain stuck in the body consciousness. In order to offer them the best possible help, a liberating funeral rite is very important. This applies to both burials and cremations.

Prayer for Funeral Rites

Beloved God of all that is, we call you in the hour of the departure of …(name) from our midst. Take …(name) to the kingdom of light and joy. Thank you that …(name) may return to you, and is free of his/her physical suffering and torment. Father-Mother-God help him through your universal love, to break away from all that he has experienced in his 'earth-clothes'. Guide him through the uncertainty and darkness that he has created for himself. We know that you never forsake a soul, and certainly never lose one. So we are certain that you are also helping …(name), and will pull him to you. We are pleased that you are taking him, and that he may come home. Now we call you …(name), because we know that you are here. Although we cannot see you, you would be amazed at what is happening now with your body. We want to help you with this ritual to break away from your earth clothes. You will now be able to recognise that you still do have a body, a spirit body of fluid light. Let go, therefore, of all thoughts of the suffering and pain of your earth body, and you will be free as a butterfly. Also let go of the as yet unresolved questions and problems from your time on earth, as they should no longer be a burden for you. Call in the divine with your hands on your heart, and feel how hot it gets in your breast. Rely on the great, divine power of love, because without it we would not be here or anywhere else. Please go back to your old, well-known reality of never-ending life. Use your powerful, pure thoughts to create new magical worlds of pure love. Be a good worker for the Creator in his infinite number of vineyards. Be blissful when you give love to others and make them happy. This we grant you from our hearts, because we love you. So we are also conscious that we must let you. To hold you back would be selfish. That would be the opposite of the unconditional universal love taught us by Jesus Christ. Our parting sorrow is a sign of our appreciation for you. But let us now look forward to your return to the pure light of divine love and your inner peace.

We now place your earthly body into the grave, so that the matter can reconnect with Mother Earth. From her, our mother Gaia, we have received our 'earth-clothes', and through her they were supported and nurtured throughout our lives.

At this point I call you mother earth and I thank you for all that you provide us for our needs and as gifts. Please take back this body-shell, and let something new be created from it.

For Ash Scattering:

In order to do this, you should call furthermore the air spirits for help:

Dear Spirit of the Elements of Air, Wind and Storms, I call you at this time. Please take the ashes of ...(name) and let them float as freely as you do, without limits and full of life forever.
Beloved God, we thank you for all the beauty that you give us here on this planet. We are especially grateful for the work of the great spirits of the Elements and of nature. And thank you for all the big things that are contained within the small. Thank you for our lives in complete freedom as your children of god. We are included in your universal love, and that is our divine heritage. We love you. Amen

21. Liberation and Clearance of Demonic Burdens and Energies

The demonic is as old as mankind, because everyone has the divine light, but also the dark and evil within him or her. Man lives in duality and that is intentional for the equilibrium on earth.

When you open a newspaper or turn on the TV, you see the bad and evil predominantly. What the politicians, bankers and church representatives are allowed and offer is beyond good and evil. I am always amazed how people take the easy way and still believe. But when a person or an institution spreads untruths for a long time, they are often based on partial truth. The devilish influences are just so clever that they help people to distribute information full of lies and slander. This happens in very finely measured doses, but always at just the right moment. It can go so far that a person doesn't even feel herself any more, or doesn't perceive it as real. Then perhaps they might not see colours anymore, or find that they cannot bear to hear cheerful music. Soon he/she acts disturbed or stares into emptiness and ultimately becomes manic-depressive. Of course this can also happen when a foreign soul or groups of astral beings attach themselves on or to a person. But that is precisely the problem, since these beings are 'burdensome' servants of the evil and demonic. When you speak with such, burdened people, you often see the flashes in their eyes, and thus directly into the essence of the devil. Many people feel that when they look into the mirror and then are scared about themselves.

In many old books from antiquity or the Middle Ages, but also from the last centuries, the devil is depicted as a monster with a grotesque face, sometimes also as dragon or lizard creatures. And it is indeed a ghastly sight every time when you meet such creatures that are attached on human beings. They appear inhuman and ugly and often stink worse than a compost heap or a dustbin. During a clearance, they swell up to the ceiling in front of me and also want to inspire fear.

Once I was in the Carnival and was sitting in the restaurant. Then there came several masked women. One of the groups was going as red devils with trident forks. They fawned around us and fooled with all the guests. Said one to me: "You, who apparently sees ghosts, you must admit that you have never seen a more beautiful devil than me. That's because they don't exist anyway!" I did not know the woman, but replied: "You really are a lovely little devil. But if you could see, as I do, the demon that is attached to your colleague, showing himself in full regalia, you would run home as if driven by a fury, and lock yourself inside. Please believe me." I could not see the expression on her face, because it was masked. And soon they where gone.

The church continues to do exorcisms. I have also described doing these exorcisms in my first book **Experiences with the Other Dimensions** Chapter 24, but in most cases it does not work at all. One should not repay evil with evil. It can try it, but it helps the demonic to be even stronger. What the devil cannot bear is love, pure love, the pure pink light of love, and red light of love from the heart of Christ! This is our strongest weapon. For those who are so heavily possessed, pray is no longer possible. They despise and mock the divine. It is indeed quite unpleasant to have to approach such a person. But out of charity, it should be done. Since these people are usually not capable of liberating themselves, you should prepare yourself in the following way: Ask the Angels and our Creator for help. Stand behind the occupied person and place both hands on their shoulders. Speak loudly and with conviction:

Liberation and Clearance

Great Creator God and beloved Saviour Jesus Christ, I call you with all your angels of truth and love, for help and support for this person. Help him to get rid of this evil demon. I ask you now for a great act of grace and restoration of the guaranteed freedom of his soul. Give him clarity about what he can do for himself, in order to be free in every way.

Now I call you, Darkness and Devil. You have deceived this person and made him subservient to you. You have covered his own light and thus destroyed his quality of life. You are the most insidious and ruthless being that exists on this earth. You are the strangest teacher and examiner. God has approved you for this task. But in the name of love and light I want you now to distance yourself from this person. Since I cannot hate you, I surround and immerse you in the pure pink light of divine love. The red light of love from the heart of Christ fills the space in your body where the heart is. So go now. You are free, and this man's soul is free as well.

Dear Father God and Saviour Jesus Christ, fill the heart and the soul of this person with the pure red light of love from your heart. Help him and protect him from further attacks by the darkness. Guide him when he is being seduced, by strengthening his will and the power of his love for you. Beloved angel, do not allow Darkness to return and continue his evil work. Give this person early warning, and support his strength and his courage. I thank you, God of Love and Light, for all the beauty and purity in this life. Amen

It is very important that this person then also clears away his old karmic patterns and the resulting elementals. Otherwise, a new attack is very possible, since the self-created negative patterns are still there and active. Trust in the divine light within yourself, and live with love for yourself and for everyone else.

22. Extraterrestrial Energies and Attachments

For many readers, this issue will probably be the biggest challenge. The vast majority of people probably think we are alone in the universe, and absolutely the greatest. That unfortunately is not so. When they asked Jesus if we were alone on this earth among all the stars, he said there were many earths like ours, and also much larger ones. On all there is life in different forms, but for many, life is only in spirit form. We humans are 'Starseed', which have been sent or descended from many different star origins. We are therefore different star races, which does not devalue us. By our diversity, we are extremely advanced and polished. And no one can argue he was reincarnated only on earth alone. We were often curious strangers on other planets. We were abused and suppressed there, or we did it to others.

The law of compensation is in effect, and always has been. No one can get around it. And for this reason, it is possible to meet people that still carry old suffering from back then. One can assume that different cultures from our own star are ahead in technology by thousands of years. For example, if a person has an artificial hip or knee and then dies, there is the question of whether he can let go of his body consciousness in the intermediate worlds or not. If not, he will in the next incarnation have the same hip or knee problems again, as he now has in this life. But many therapists tell me that they had patients with extraterrestrial implants, which worked independently. These are a kind of human-robot hybrids that are sometimes controlled by external entities. Most of these poor people do not have their own senses and perceive their own body as foreign. For the doctors these patients are outside of any normal understanding, and are not treatable. I myself have been really challenged, like many other therapists, when I have met such unfortunates. For these people, I have developed a clearance prayer and ritual, which could help in many cases.

Clearance Prayer for Manipulation by Extraterrestrials

I call you, beloved creator God, and you, Saviour Christ, for help and support for me (or these people here). *Help me restore the purity, the integrity, as well as the freedom of my soul that you have guaranteed.*
Help me to liberate myself from these manipulative implants that were attached to me. Or I must now accept them, because I have used them to enslave others. Help me to find those souls I have made unhappy and have caused to suffer. Please forgive me my very large

debt and let me be free. Help me to address these beings that are imprisoning me and manipulating me. I forgive you for what you have done.

I surround myself in the most powerful light of Christ-love. I am free. You are free. God and dear Jesus Christ, help me and them, to close the wounds of our old suffering. And make their body and also mine as perfect as it was when you gave it to us. I love you, Father-Mother-God and Jesus Christ in me and in all that is. So be it. I thank the eternal light of love. Amen

23. Implants From Previous Lives

Many children are born with faults and weaknesses in their organs, but also in their musculoskeletal system. Of course, many adults also carry inexplicable suffering on earth. They are actually healthy, but suffer from pseudo-pain or phantom illness. No doctor has a suitable explanation or clue of what the patient is bringing to him. These pains are real, but are not visible on an ultrasound device or on an X-ray machine. This ongoing suffering can not be detected even with a CT scanner. It is not easy for doctors who have such patients. They want to help them but do not know how. Such people are then passed on for examination by one specialist to another and back again. Many become like a hot potato that no one wants to burn their finger on. After a while everyone gets a diagnosis for their money: 'psychosomatic disorder' or in German 'eingebildete Krankheit' (imaginary illness). Not infrequently, the patient receives psychotropic drugs to calm them down. Mostly the pain and suffering continue to exist, but people suddenly think that they are crazy, or are about to begin being crazy. They think, "Yes, now I'm incurable". This setting creates a new Disease-Elemental, that will not go away by itself. It accompanies us back to the next life.

When such a sufferer comes to me, I ask his soul if he had surgery in an earlier life • whether he received an organ • whether he had a pacemaker • whether he had artificial joints e.g. hip, knee, other joints or tooth implants.

If something like that was previously in his body, a disturbance is likely, even if he tolerated the implants well. Now, in this life, in this body, the former third-party product or donor organ becomes a problem. The memory of the foreign matter is a disturbance energy that is still active. This disrupts the flow of life for the body-spirit in this body, now. Such an energy blockade can occur even in formerly amputated limbs.

It is best to speak with the guiding spirits of your body and together with them, remove such old faults with the help of prayers (Chapter 18.1).

Prayer for Liberation from Old Diseases

I call you great spirit of my body. I call you, countless spirits in all my cells. I now speak to you, guiding spirits of all my organs. Please listen to me carefully.

You work for me day and night without interruption, without me knowing exactly what you do. But I am suf-

fering, because you are suffering and telling me that something old still bothers you. I ask you to erase all memories of past lives and suffering that have happened to you. Erase the information and energy vibrations of artificial implants and foreign bodies from your midst. Erase the memory and suffering from other foreign organs. Delete all commands from me that you made the sick. Now clear anything that is against your actual health and well-being. That's what I want.

Beloved body remind yourself of your basic function, your prime directive, namely well-being. You are whole and complete.

I thank you for your great and important task. I love you and I apologise to you for all that I subconsciously worked against you and have done to you.

I call and beg you great Creator Spirit of my body and of all that is, please help my body through your love and with the Angel of Healing. I am healed! So be it. I thank you and love you, because you create perfection in me and have given it to me. I give my thanks and my love to you Prime Creator and to the great light of your ever-present love. Amen

Epilogue

We received an important announcement from Archangel Gabriel in December 2011. He asks that we should give lectures or seminars to inform people about humanity's **collective** suffering. He says the following:

There is a large network of suffering energy surrounding the globe. All mankind are caught in it. On TV we see an accident with fatalities and injuries in India or we hear something about a mudslide in Colombia with hundreds of dead, or a flood in Thailand, where many people were drowned, etc. We all suffer because of these victims. They make us feel so sorry that we suffer physically.

What we don't realise is that by reacting like this, we are exacerbating the global pain and suffering. In this way, more people are oppressed with sadness and pain.

Gabriel thinks it would be a great advantage if we would keep ourselves completely neutral when we receive such negative messages, and even better, visualize the pure divine light of love and send it there.

Ritual from Archangel Gabriel

When we encounter such suffering we should immediately ensure that it is not retained within us and that we do not connect with the patterns of suffering and the negative energies of ourselves or others. The following sentence is useful for this.

Via my soul, I am always connected with divine universal love, and never with the patterns and themes of suffering or negative energy from within me or from other people.

So suffering can no longer be recalled within us, nor reinforced further. The divine is always perfect and whole. God has never created diseases or suffering, we did that, his children.

Anton Styger, Styger-Verlag

www.styger-verlag.ch
www.antonstyger.ch

Spiritual Realms on Earth

Element-spirit realms
Earth and crystal worlds
Fire and light spirits
Air and wind spirits
Water, sources, lakes and seas
Plants above ground, under water
Animals in the air, in the ground, in the water
    on micro level, on macro level

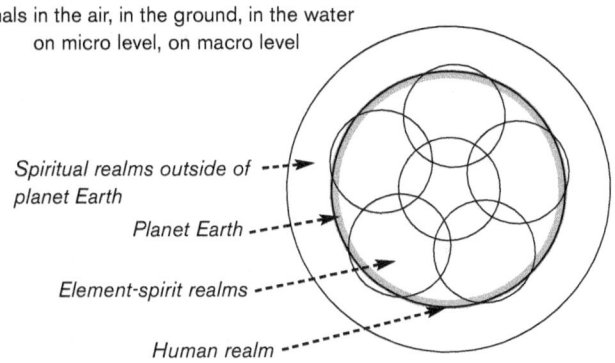

Spiritual realms outside of planet Earth
Planet Earth
Element-spirit realms
Human realm

Spiritual Realms around the Human Realm:

Nature spirits
Earth matter
Earth spirits
Nature spirits
Devas
Gnomes
Dwarfs

Plant spirits
Elves for the trees
Fairies for the flowers

Animal souls directed
above ground and air
underground
micro level and macro level
water animals

Disturbing souls and energies
Earth-bound souls
Astral beings of the deceased
Negative waste of astral beings
Demonic energies

Anton Styger's book: Volume 1
Pages: 474
ISBN: 9781326529192   and   978-1-326-52919-2
Ordering: **www.lulu.com**

Soul Liberation
Subtitle: **Experiences with Other Dimensions**

Have you ever wondered why you feel powerless and exhausted after visiting an event or just a shopping spree in town? Are you blaming the lethargy of daily conditions on general stress, uneasiness, lack of sleep or similar conditions? Do you sometimes become aware that you can see, hear or feel things that others cannot perceive? In this book you find out, how people can be freed from heavy burden and energy robbers. Bound souls, astral beings or demonic energies exist even if affected people do not know or believe that there is such a thing.

The central components of the attitude of life of the medial author are the faith in rebirth and personal responsibility. Based on his many experiences, you can see how versatile and great the spiritual world is and also acts.

The instructions for prayers on liberation- and protection are tested practically which you can use in your everyday life and will help you to enjoy the lightness of being.

Anton Styger's Buch: Band 1
536 Seiten ISBN: 978-3-033-01586-9
Erlebnisse mit den Zwischenwelten
Untertitel: **Seelenbefreiungen**

Haben Sie sich schon mal gefragt, warum Sie sich nach einem Besuch, nach einer Veranstaltung oder einfach einem Einkaufsbummel in der Stadt völlig kraftlos und ausgelaugt fühlen?
Schieben Sie die Energielosigkeit der Tagesform, allgemeinem Stress oder Unwohlsein, zu wenig geschlafen oder ähnlichen Erklärungen zu? Haben Sie manchmal das Gefühl, Sie können Dinge sehen, fühlen oder hören, die andere nicht wahrnehmen können? In diesem Buch erfahren Sie, wie schwer belastete oder besetzte Menschen von Energieräubern befreit werden. Gebundene Seelen, Astralwesen oder dämonische Energien existieren, auch wenn die befallenen Menschen nicht gewusst oder geglaubt haben, dass es so etwas gibt.

Der Glaube an Wiedergeburt und Eigenverantwortung sind zentrale Bestandteile der Lebensauffassung des medialen Autors.
Anhand seiner vielen Erlebnisse können Sie erkennen, wie vielseitig und grossartig die geistige Welt ist und auch wirkt.

Die Anleitungen, Befreiungs- und Schutzgebete können Ihnen helfen, Ihren Alltag unbeschwert zu geniessen.

Oraciones para el Alma   de Antón Styger
ISBN: 978-612-46362-0-2   173 Páginas
Ordenar: **www.styger-verlag.ch**

¡Toma la vida nuevamente en tus manos!

Este libro revive en nuestra comprensión de Dios. Antón Styger vive en Ägerital, Suiza. Es autor de la serie de libros denominada Vivencias con los mundos intermedios, y se dedica a sondear casa y establos donde la gente y los animales sufren a causa de la presencia de zonas con perturbaciones geománticas o electrotécnicas. Además, casi siempre se ha enfrentado también con cargas perturbadoras ajenas muy sutiles. Pero, clarividente desde su niñez, está bastante familiarizado con este fenómeno. Así, durante décadas ha desarrollado diversos ejercicios y oraciones para liberarse tanto de esas energías ajenas e invisibles como de patrones del alma muy antiguos y limitadores, producto frecuentemente de heridas recurrentes.

Motivado por da especial resonancia de sus primeras dos obras, el autor nos entrega ahora. Resumidos en un manual, las instrucciones y los rituales consignados en aquellas. Se trata en realidad de oraciones parecidas a diálogos amorosos con el Creador. No es casual, por cierto, que un sinnúmero de lectores haya utilizado exitosamente, con alegría y dedicación, sus ejercicios e invocaciones.

Anton Styger's Taschenbuch: **Gebete für die Seele**
ISBN: 978-3-9523755-7-0
180 Seiten

Das Leben wieder selbst in die Hand nehmen!
Dieses Buch wird Ihre Suche nach dem eigenen Ich beleben und
Ihr Gottverständnis positiv beeinflussen.

Anton Styger lebt in der Schweiz im Ägerital. Er ist der Autor der
Buchreihe **Erlebnisse mit den Zwischenwelten** und befasst sich
mit dem Vermessen von Häusern und Ställen, dort wo Menschen
oder Tiere unter geomantischen oder elektrotechnischen Störzonen
leiden. Dabei stösst er fast immer auch auf störende Fremd-
belastungen feinstofflicher Art. Von Kindesalter an entwickelte er
zahlreiche Übungen und Gebete um sich von ihnen zu befreien.

Aufgrund der Resonanz auf seine inzwischen 5 Bücher fasste der Autor die Anleitungen und Rituale zu einem Handbuch zusammen.

Es ist angefüllt mit den gesammelten Gebeten des Autors, die sich wie Liebesdialoge mit dem Schöpfer lesen. Seine Übungen und Anrufungen werden bereits von unzähligen Lesern mit Freude und Hingabe erfolgreich angewendet.

www.ingramcontent.com/pod-product-compliance
Lightning Source LLC
Chambersburg PA
CBHW070231180526
45158CB00001BA/388